HUSSERL AND THE SCIENCES

SCIENCES

SELECTED PERSPECTIVES

The Philosophica series covers works dealing with perennial questions of the history of philosophy. The series particularly seeks works written within the European Continental and the analytic traditions. In conformity with the Press's editorial policy, the series welcomes manuscripts written in either English or French.

Series Director: Josiane Boulad-Ayoub

φ PHILOSOPHICA 55

HUSSERL AND THE SCIENCES

SELECTED PERSPECTIVES

EDITED BY RICHARD FEIST

UNIVERSITY OF OTTAWA PRESS

This book has been published with the help of a grant from the Canadian Federation for the Humanities and Social Sciences, through the Aid to Scholarly Publications Programme, using funds provided by the Social Sciences and Humanities Research Council of Canada.

University of Ottawa Press gratefully acknowledges the support extended to its publishing programme by the Canada Council and the University of Ottawa.

We acknowledge the financial support of the Government of Canada through the Book Publishing Industry Development Program (BPIDP) for our publishing activities.

National Library of Canada Cataloguing in Publication

Husserl and the sciences : selected perspectives / edited by Richard Feist.

(Philosophica; 55)
ISBN 0-7766-3026-1

1. Phenomenology. 2. Husserl, Edmund, 1859–1938. 3. Philosophy and science. I. Feist, Richard, 1964– II. Series: Collection Philosophica; 55.

| B829.5.H87 | 2003 | 142'.7 | C2003-904066-6 |

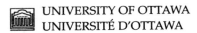

UNIVERSITY OF OTTAWA
UNIVERSITÉ D'OTTAWA

Copy editing: Carol Tobin

All rights reserved. No parts of this publication may be reproduced or transmitted in any form or by any means, electronic or mechanical, including photocopy, recording, or any information storage and retrieval system, without permission in writing from the publisher.

ISBN 0-7766-3026-1
ISSN 1480-4670
© University of Ottawa Press, 2004
   542 King Edward, Ottawa, ON Canada K1N 6N5
   press@uottawa.ca   www.uopress.uottawa.ca

Printed and bound in Canada

B
829.5
H828
2004

090806-3162 P8

CONTENTS

———

———

CONTENTS

ACKNOWLEDGEMENTS

I would like to thank the Social Sciences and Humanities Research Council of Canada as well as the Aid to Scholarly Publications Programme for their funding of the publication of these papers.

I would like to thank the Austrian Embassy and the Royal Netherlands Embassy for their financial assistance. I especially thank His Excellency, Wendelin Ettmayer, the Austrian Ambassador, for his hospitality and interest.

I would like to thank Carol Tobin for her superb editing work and Lynne Mackay for all her assistance. I also would like to thank Vicki Bennett who has been with this project from the beginning and has been helpful and patient throughout.

I also would like to thank Anoop Gupta for all his help.

Finally, a thanks to my wife Barbara and son Henri, to whom I dedicate this work.

ABOUT THE AUTHORS

John Bell is professor of philosophy at the University of Western Ontario. He is the author of several texts and articles in mathematics as well as articles on the philosophy of mathematics.

R. Philip Buckley is professor of philosophy at McGill University. He is the author of _Husserl, Heidegger and the Crisis of Philosophical Responsibility_ as well as numerous articles on phenomenological analyses of the self and on political philosophy. Forthcoming is his translation of a volume of Husserl's _Essays on Culture_ to appear in the series _Edmund Husserl: Collected Works_.

Richard Feist is professor of philosophy at St. Paul University. He is the author of numerous articles on logic, mathematics, and phenomenology. Currently he is at work on a book about Husserl and Weyl.

Denis Fisette is professor of philosophy at l'Université du Québec à Montréal. He is the author of _Lecture frégéenne de la phénoménologie_ and co-author of _Philosophie de l'esprit_. He has published numerous articles on phenomenology and edited the forthcoming _Husserl's 'Logical Investigations' Reconsidered_. He is currently working on a book about phenomenological psychology.

ABOUT THE AUTHORS

Yvon Gauthier is professor of logic and the foundations of mathematics at the University of Montreal. He is the author of numerous books, the most recent being *Internal Logic: Foundations of Mathematics from Kronecker to Hilbert*. He has published numerous articles and is currently working on a book *The Content of Logic: Arithmetic and its Extensions*.

Pierre Kerszberg is professor of philosophy at the University of Toulouse. He is the author of *The Invented Universe, Critique and Totality*, and *Kant et la Nature* as well as numerous articles. His work deals with both the philosophy of science, with a special emphasis on cosmology, and the philosophical tradition issuing from Kant and phenomenology.

Ulrich Majer is professor of philosophy at the University of Göttingen. He is the author of several articles on phenomenology and on the history and philosophy of mathematics.

Mathieu Marion currently holds the Canada Research Chair in the Philosophy of Logic and Mathematics at l'Université du Québec à Montréal. He is the author of *Wittgenstein, Finitism, and the Foundations of Mathematics*.

Herman Philipse is professor of philosophy at the University of Utrecht. He is the author of several books including *Heidegger's Philosophy of Being*. He has published numerous articles on epistemology, metaphysics, phenomenology, and the history of philosophy.

Sonja Rinofner-Kreidl is professor of philosophy at the University of Graz. She is the author of several articles on Husserl's philosophy and the phenomenological tradition.

INTRODUCTION

Richard Feist

The founder of the phenomenological movement, Edmund Husserl (1859–1938), lived through a dynamic time for the sciences.[1] Not only were there major developments in mathematics and physics, but some of the greatest practitioners of these disciplines were pursuing foundational questions with an unprecedented depth and rigour. Although Husserl did not directly contribute to these developments, it is not correct to say that he simply sat on the sidelines. He personally knew and corresponded with several of the finest scientific and mathematical minds of the time. It is, therefore, not surprising that the relationship between Husserl's philosophy and the sciences is long and complicated. Currently, there is a growing interest in this relationship and the following collection of essays is intended to add to the discussion.

As is well known, Husserl's initial interest was in mathematics. From 1878 to 1881 at the prestigious University of Berlin, he studied Karl Weierstrass's work in analysis and Leopold Kronecker's in number theory.[2] Soon he departed for Vienna University, where in 1883 he completed his doctoral thesis in the theory of variations. As was then typical, Husserl had to confirm his doctoral work by defending a "thesis of habilitation" before an academic tribunal. Husserl defended his thesis, "The Concept of Number," at the beginning of his time (1866–1900) at the University of Halle. One of the members of his tribunal was Georg Cantor, the founder of transfinite mathematics and one of the most important mathematicians of the nineteenth century, with whom Husserl developed a lasting friendship. *The Philosophy of Arithmetic* appeared in 1891 and is arguably Husserl's first major work. It indeed bears the influence of Cantor.[3]

Shortly afterwards, Husserl corresponded with the Jena mathematician, Gottlob Frege, who is often regarded as the founder of analytic philosophy. Like Cantor, Frege had a large influence on Husserl; however, the precise nature of this influence remains a source of debate.[4]

Husserl then worked for a number of years (1901–1916) at the University of Göttingen, then one of the greatest mathematical research institutes in the world. These years were witness to several major works, most notably *Logical Investigations* (1900–1901) and the first volume of *Ideas* (1913). Husserl regarded *Logical Investigations* as the "breakthrough to phenomenology." The controversial *Ideas* is often held to mark Husserl's so-called 'transcendental turn.'[5] Both of these texts bear the strong influence of the type of mathematics that was being conducted in Göttingen by mathematicians such as Felix Klein and David Hilbert, the latter being one of the most important mathematicians of the twentieth century.[6]

The Göttingen milieu also exposed Husserl to the latest developments in physics. During this time Einstein published his works on relativity. Göttingen thinkers rapidly absorbed the new views of space and time and analyzed their foundations. It was the Göttingen mathematician, Hermann Minkowski, who translated the theory of Special Relativity into the geometric form in which it is taught to this day. Hilbert and Klein also worked in the mathematical foundations of relativity theory. But perhaps most important were the labours of Hermann Weyl. Weyl attempted to develop the Riemannian geometric foundations of General Relativity such that it would extend beyond the borders of gravitation to include electro-magnetism.[7] There were many critics of his approach and it was largely abandoned, although it is currently enjoying a renaissance. What is relevant for us here is that Weyl claimed that Husserl's thought influenced his own investigations into the Riemannian foundations of Relativity as well as his investigations into the foundations of analysis.[8] Some time later Kurt Gödel was to claim that Husserl's phenomenology was the means by which we should interpret the incompleteness theorems.[9]

After Göttingen were the final Freiburg years, 1916–1939. It could be said that during this time Husserl deepened his insight into his previous works rather than developed new ideas.[10] Themes such as history and society started to play a larger role in Husserl's thought than they had previously.[11] In effect, we should not understand Husserl as solely concerned with the foundations of mathematics and science; rather, we should understand him as embedding the foundations of mathematics and the sciences within the broader context of a general philosophy. Over the years this general philosophy took on different forms, beginning with an attack on the radical specialization and isolation of disciplines from each other to the notion of all disciplines having their roots in the 'lifeworld.' Husserl laboured to find a general philosophy that could account for the gamut of human experience, which includes those that involve foundational problems.[12]

Herman Philipse begins our series of papers by setting the epistemological stage. In "Edmund Husserl and the History of Classical Foundationalism," Philipse argues that Husserl's work is one of the final attempts within the research programme of "classical foundationalism." Philipse outlines the origins of this programme in Aristotle's thought and discusses its developments throughout the subsequent centuries until its decline during the time of Husserl.

Sonja Rinofner-Kreidl's paper, "What Is Wrong With Naturalizing Epistemology? A Phenomenologist's Reply," explores a general epistemological disagreement regarding the foundations of our knowledge of the world by contrasting the different approaches of Husserl and Quine. Rinofner-Kreidl focuses on the issues of circularity, foundationalism, and skepticism. She argues that Husserl's phenomenological programme, with its descriptive apriorism, is able to offer a viable alternative to Quine's naturalist programme.

Retaining the theme of an epistemological clash between Husserl and another major philosopher, Denis Fisette, in "*Erläuterungen*: Logical Analysis vs. Phenomenological Descriptions," articulates what is at stake in the Husserl–Frege debate regarding the epistemological status of the fundamental concepts of mathematics. Fisette contends that this debate is not over an isolated, technical corner of the philosophy of mathematics; rather, it involves a deep epistemological disagreement over the limits to which one can analyse fundamental concepts in general.

Ulrich Majer pursues the epistemological clash regarding the foundations of mathematics by focusing on the basic concepts of geometry. In "Husserl and Hilbert on Geometry," Majer examines Klein's group-theoretic and Hilbert's axiomatic approaches to the foundations of geometry; both belonged to the Göttingen milieu. Majer argues that these differ regarding the nature of geometry's founding intuition. Majer also argues that Husserl's view of intuition is closer to Hilbert's than Klein's.

The historical dimension of Husserl's thought is then explored with respect to mathematics. Yvon Gauthier, in "Husserl and the Theory of Multiplicities '*Mannigfaltikeitslehre*,'" discusses Husserl's theory of multiplicities, which was a recurrent and unchanging theme throughout his life's work. Gauthier argues that the main source of this theory is the late nineteenth century project of "general arithmetic," whose key proponent was the Berlin mathematician Leopold Kronecker.

Intuition is a slippery term in philosophy. It is often used inconsistently and is rarely explained. The problem of the nature of founding intuitions, which was discussed by Majer, is then revisited by Mathieu Marion in "Husserl's Legacy in the Philosophy of Mathematics: From Realism to Predicativism." Marion discusses how several different philosophical positions claim to be based on phenomenology before he concentrates on two: the predicativism of Hermann Weyl and the realism of Kurt Gödel. Marion argues that although neither of them embraces a view of mathematical intuition *exactly* like Husserl's, Gödel's is closer than Weyl's.

The problem of intuition is continued in Richard Feist's paper "Husserl and Weyl: Phenomenology, Mathematics, and Physics." Feist also compares Weyl and Husserl on the nature of founding intuitions in mathematics. Feist then relates this discussion to Weyl's generalization of the Riemannian underpinnings of the General Theory of Relativity.

Finally, John Bell carries the discussion of intuition and the Weyl–Husserl connection in a different direction in "Herman Weyl's Later Philosophical Views: His Divergence from Husserl." Bell argues that Weyl eventually came to the realization that Husserl's philosophy lacked proper treatment of the problem of the external world and the nature of the self. Regarding the external world, Weyl moved towards the philosophy of Ernst Cassirer. Regarding the self, Weyl moved towards the philosophy of existentialism as well as religious mysticism.

Husserl attempted to locate the problems of the foundations of the sciences within a broader philosophy. Pierre Kerszberg, in "From the Lifeworld to the Exact Sciences and Back," argues that the scientific object, according to Husserl, must ultimately be grounded within the 'lifeworld.' The problem is that the objects of physics are highly abstract and seem to have no relationship to experience. Kerszberg argues that according to Husserl, the scientific object is ultimately grounded within the "hidden dimension of the body." But this hidden dimension is not at all accessible to a Cartesian solipsistic consciousness; rather, it is opened and revealed to consciousness only through intersubjective experience. In other words, the scientific object finds its proper phenomenological foundations within a community.

The themes of community and science are then continued by Philip Buckley in "Husserl on the Communal Praxis of Science." Buckley argues that Husserl's thought contains the seeds of a social philosophy. It is interesting to note that the current interest in Husserl's reflections on society is small, but growing. Buckley stresses that there is a continuity to Husserl's work. Many themes, such as 'history,' which are held to be later products of Husserl's thought, can be found in implicit form in the early works. But as Buckley argues, this implicit social philosophy is modeled on that of an actual subsociety, namely, the society of mathematicians.

It is certainly not the case that these essays cover the entirety of Husserl's thought in relation to the sciences. They are only a series of perspectives, intended to contribute to the growing interest in Husserl's philosophy and the sciences. Moreover, there are cases in which the authors disagree on the interpretation of Husserl's thought. There are disagreements as to Husserl's place in the history of epistemology. Should he be regarded as the last of the classical epistemologists or as the author of an entirely new approach to epistemology? Moreover, what exactly is Husserl's notion of a fundamental intuition and how does it relate to the views expressed by the mathematicians of the Göttingen milieu? Nonetheless, it is through the medium of debate, with the resulting conflicts – and agreements – arising out of these perspectives that a fuller picture of Husserl's philosophy in relation to the sciences will eventually emerge.

Notes

1 For more bibliographic information see J.N. Mohanty, "The Development of Husserl's Thought," in *The Cambridge Companion to Husserl*, ed. Barry Smith and David Woodruff Smith, 45–77 (Cambridge: Cambridge University Press, 1995). For more on Husserl's epistemology and its place in western thought see Rinofner's and Philipse's papers in this volume.

2 For more on this see Gauthier's paper in this volume.

3 For more on the Cantor–Husserl connection see Claire Ortiz Hill, "Frege's Attack on Husserl and Cantor," *The Monist* 3 (1994): 345–77.

4 For more on this see Fisette's paper in this volume.

5 See Mohanty, "The Development of Husserl's Thought."

6 See Majer's paper in this volume.

7 See Feist's paper in this volume.

8 For more on Weyl's difficult relationship with Husserl see Bell's paper in this volume.

9 See Marion's paper in this volume.

10 Mohanty, "The Development of Husserl's Thought," 63.

11 See Buckley's paper in this volume.

12 See Kerszberg's paper in this volume.

PART I

PHENOMENOLOGY, EPISTEMOLOGY, AND THE SCIENCES

CHAPTER ONE

Herman Philipse

EDMUND HUSSERL AND THE HISTORY OF CLASSICAL FOUNDATIONALISM

———

1. INTRODUCTION: PROBLEMS AND METHOD

According to many present-day epistemologists, the justification of scientific theories is relative in at least two respects. Whether a specific theory is justified at time *t* depends both upon the set E of empirical data available at *t*, and upon the set R of rival theories which are considered by the relevant scientific community at that time. Indeed, a theory is justified at time *t* if and only if it performs better than its rivals in terms of the accepted criteria for theory choice, and one decisive criterion for theory choice is some version of the criterion of empirical superiority. As a consequence, a theory that is now justified might cease to be justified in the future because of two reasons. New empirical data may top the balance in favour of an existing rival theory, or a new rival theory may be designed that performs better.

This relative notion of the justification of theories by competition is not only a rough model of justification, it also solves the traditional problem of demarcation. We do not need to look for an atemporal criterion of demarcation that enables us to decide whether theories are scientific or not (such as the criterion of falsifiability) for we now have a time-relative criterion. *Theories* are not scientific or unscientific in the abstract, but *someone* has an

———

unscientific *attitude* if he or she prefers a given theory T at time *t* even though at that time there is another theory available that is clearly superior to T in terms of the criteria for theory choice.[1]

The model of justification by competition must be applied also at the meta-level of justifying epistemological theories. Like scientific theories, epistemological theories and models are not invented and evaluated in the abstract. They are typically designed to solve problems that arise in determinate historical circumstances, for example because new developments in the sciences conflict with the epistemological *status quo*. The criteria for theory choice in the domain of empirical science have their analogues in the domain of epistemology, although here, of course, the data set does not consist of natural phenomena; rather, it is the set K of (scientific) knowledge at a given time *t*. Normative epistemological theories should be empirically adequate in the sense that they enable us to provide a rational reconstruction of paradigmatic scientific developments. An epistemological theory, then, is justified at time *t* if it provides a rational reconstruction of the scientific developments up to *t* that is superior to the reconstructions provided by its rivals. This meta-epistemology of epistemologies explains the co-evolution of science and epistemology.

In the tradition of analytical philosophy, the rules of academic etiquette seem to exclude an application of this model of justification by competition on the meta-level, since the model prescribes that in trying to understand and evaluate epistemologies, specific historical circumstances must be taken into account. Analytical philosophers, however, typically assume that it is imperative to discuss the major epistemological models proposed in the past, such as foundationalism and coherentism, without reference to the historical contexts in which these models were developed as if the evaluation of epistemologies could proceed in an intellectual vacuum.[2] In my view, this a-historical style of analytical epistemology accounts for both the characteristic barrenness of the discipline and its irrelevance to scientists and the history of science. Moreover, in many cases analytical presentations of epistemological models such as foundationalism are distorted, because authors abstract from the time-bound reasons for developing these models in specific historical circumstances, and omit to discuss features that were once deemed essential.

In this paper I want to explore by means of a case study how these defects may be remedied, focusing mainly on the example of Edmund Husserl's foundationalism. I shall attempt to answer three questions. First, which type of foundationalism did Husserl adhere to? Second, was he justified at the time in endorsing his type of foundationalism as a general theory of science? Finally, would one currently be justified in accepting Husserl's foundationalism? I argue that Husserl's foundationalism is a version of what might be called 'classical foundationalism,' so named because it ultimately derives from Aristotle. A succinct rational reconstruction of the history of epistemology from Aristotle to Husserl will show that classical foundationalism is a research programme whose development was determined not only by an internal logic but also by the scientific context. During Husserl's lifetime, the programme degenerated, and at the end of this paper I consider a section of Heidegger's *Sein und Zeit* which shows that the programme could not cope with the scientific revolutions of the early twentieth century.[3]

2. THE ORIGINS OF CLASSICAL FOUNDATIONALISM

In the literature there are various definitions of foundationalism, either as a theory of knowledge or as a theory of justification. These definitions have two points in common. All varieties of foundationalism require a distinction, among beliefs that count as justified or as knowledge, between those that are basic and those that are derived. Furthermore, foundationalist theories hold that justification is predominantly one-directional, that is, from basic beliefs to derived beliefs, and that justification of non-basic beliefs is a logical derivation of some kind.

Within the broad class of foundationalist theories, so defined, many different types of foundationalism may be distinguished, depending upon the answers that these theories provide to three questions:

1) which kinds of beliefs are considered to be 'basic'?
2) how are these basic beliefs justified?
3) how is the logic of derivation to be characterized?

Sometimes, the honorary title of 'classical foundationalism' is reserved for the doctrine that basic beliefs are beliefs that concern the nature of our own sensory states, our own immediate experience.[4] But from a historical point of view, this terminology is unfortunate. Both British empiricism and some logical positivists who held this view represent stages in the development of a much older research programme in epistemology, a programme inaugurated by Plato and Aristotle. Instead, the *epitheton ornans* 'classical' should be linked to a feature common to all or most phases in the development of this programme. These phases may then be called 'classical' because they were in accordance with the ancient model of foundationalism constructed by Aristotle.[5]

Within the context of Greek philosophy, foundationalism was developed as an answer to two problems that were pressing at the time: (a) how is knowledge (*episteme*) to be characterized in contradistinction to mere opinion (*doxa*), and (b) how can one show that knowledge is possible at all? Plato's main dialogue on this issue, *Theaetetus*, ends in an *aporia* with regard to (a). Socrates had argued that knowledge should not be defined as mere true judgment, because judgments made upon hearsay cannot count as knowledge even when they are true. With his characteristic irony, Plato now makes Theaetetus suddenly remember (201c/d) what he once heard a man say, to wit, that knowledge is true judgment with an account *logos*. But this theory leads into insuperable difficulties. Having an account might mean, for instance, that one knows an object O if one is able to go through its elements. This interpretation either leads to a *regressus ad infinitum* (in order to know the elements, one has to go through the elements of these elements, etc.) or one should conclude that one cannot know the elements, which is absurd, for how could one know the whole without knowing the elements? Having an account might also mean being able to tell the difference between the object O and other objects. This second interpretation makes the definition of knowledge circular, for how could one tell the difference if one does not *know* this difference?

We may suppose that these final sections of *Theaetetus* inspired Aristotle's epistemological research programme. In *Prior Analytics*, Aristotle had developed his deductive logic. On the basis of his theory of syllogistics, he could explain in *Posterior Analytics* what it is to have an 'account' *logos*. In chapter A2, Aristotle defines "knowledge" as a true judgment with an account, and he explains what it is to have an account by saying that an account is a scientific deduction of the true judgment, that is, a deductive proof the premises of which are explanatory with regard to the conclusion (71b, 17–19). It may seem that the requirement of deductive proof is extravagant because we now think that it cannot be met in the empirical sciences. Yet the requirement is plausible given Aristotle's common-sensical approach to knowledge. It is part of the logical grammar of the verb 'to know' (and its Greek counterparts) that propositions of the form 'X knows that p' imply that p is true; this is why truth is made into a requirement for knowledge. And if X can have no knowledge unless his or her opinion is true, it is plausible to assume that X's *claim* to knowledge must be warranted by an account which *guarantees* the truth of what X claims to know, that is, by a proof.[6]

Aristotle's definition of knowledge as belief or judgment proven to be true by an explanatory deduction answers question (a), but it immediately raises the sceptical question (b): how can one show that knowledge as defined is possible at all? In particular, how is one able to avoid the objections of regression and circularity which Socrates raised in *Theaetetus*? It is no accident that Aristotle discusses these objections in chapter A3 of *Posterior Analytics*, and he proposes what I shall call 'classical foundationalism' as an answer to these objections. The problem with which Aristotle deals here has been baptized, instructively, the Münchhausen trilemma.[7] If knowledge that p is belief that p, proven to be true, we cannot know that p unless we also know that the premises of the proof that p are true, which we cannot know unless we also know that the premises of the proofs that the premises of the proof that p are true, are true, and so on. Aristotle admits that knowledge would be impossible if this regress goes on indefinitely, "for it is impossible to go through indefinitely many things."[8] We might avoid the regress by admitting circular deductions, but these do not amount to proofs: they beg the question. Aristotle's solution consists in claiming that there are basic beliefs or "first principles" that are known to be true without proof. Accordingly, the definition of knowledge as belief proven to be true cannot apply to them, and Aristotle calls knowledge of the first principles *sofia* instead of *episteme*. A deductivist variety of foundationalism is the result of these considerations, and it is this deductivist variety which I call 'classical foundationalism.'

Classical foundationalism is best conceived as a research programme, for although it provides a definite answer to 3) by characterizing the mode of derivation of derived knowledge as deductive proof, it leaves open issues 1) which kinds of beliefs are considered to be 'basic' and 2) how these basic beliefs are justified. Let me call the complex issue (1 & 2) the 'problem of the first principles' in classical foundationalism. The history of Western epistemology from Aristotle to Husserl and Heidegger may be reconstructed, to a large extent, as the development of the research programme of classical foundationalism which has been the dominating epistemological programme for more than two millennia. Like his younger contemporary Euclid, Aristotle distinguished two types of first principles: principles common to all the sciences and principles specific to a particular discipline, such as astronomy or physics. It would be the task of philosophy to lay the foundations of all the special sciences by establishing the common first principles, and this is why Aristotle (and Descartes) used the expression 'first philosophy.' Of course philosophy also had to solve problem 2) both for these general principles and for the principles of the special sciences. Aristotle solved this latter problem in chapter B19 of *Posterior Analytics*.

In order to understand Aristotle's solution, one must grasp the specific form that the problem of the (specific) first principles acquired within the framework of classical foundationalism. In the sciences, we have to deduce general theories from first principles. Aristotle held that 'basic' knowledge in the empirical sciences must be justified by observation, and that observation is of individuals. Yet universal theories cannot be deduced from singular judgments of observation. Hence, the first principles of the sciences must be universal too, and Aristotle claimed that the logical form of scientific proofs is what the scholastics called *Barbara*. Moreover, scientific proofs cannot establish the truth of theories unless our knowledge of the first principles is secure. For this reason, Aristotle required that the first principles are necessarily true and that we are able to grasp these necessary truths. The problem which Aristotle had to solve may now be formulated as a paradox: classical (deductivist) foundationalism seems to exclude empiricism, because empiricism appears to imply that the first principles of the sciences are singular judgments of observation, whereas classical foundationalism requires universal first principles that are necessarily true. How are we to combine classical foundationalism with empiricism? How can observation justify first principles that are necessarily true and universal?

This is the problem Aristotle set out to solve in chapter B19 of *Posterior Analytics*, and he solved it by an idiosyncratic notion of induction (*epagoge*). According to this notion, a repeated observation of individuals belonging to a natural kind will yield 'experience,' that is, an observational memory of these individuals. And experience yields intuition of a universal "that has come to rest in the mind" (100a, 7–8). In other words, it was Aristotle's notion of individuals as "*concreta*" consisting of a general essence and individualizing matter, that enabled him to solve the problem of the first principles: experience gives us access to universal essences, and the absorption of essences in the mind on the basis of experience justifies universal first principles of the empirical sciences that are necessarily true of all individuals of a natural kind. Let me call this solution *Aristotelian* classical foundationalism. Aristotle succeeded in combining empiricism with deductivist foundationalism by postulating an intuition of essences based upon experience. I shall argue that Husserl endorsed classical foundationalism in the *Prolegomena* to his *Logische Untersuchungen* (*Logical Investigations*, 1900) and that, somewhere between 1900 and 1913, he came to see that he could rescue classical foundationalism only by endorsing its Aristotelian variety. However, in order to assess whether in 1913 Husserl was justified in accepting Aristotelian classical foundationalism, we have to survey the development of classical foundationalism as a research programme from the scientific revolution to the end of the nineteenth century.

3. RESEARCH PROGRAMME OF CLASSICAL FOUNDATIONALISM AND THE SCIENTIFIC REVOLUTION

From the scientific revolution in the seventeenth century onwards, the fate of classical foundationalism has been closely linked to developments in the sciences. New solutions to the problem of the first principles were proposed whenever the advance of science refuted the received view. Although attempts were made by empiricists such as Hume to replace classical foundationalism by an alternative research programme, these alternatives did not win over the philosophical community, and classical foundationalism remained the dominant epistemological programme until the interbellum in the twentieth century. Sections 3–5 contain a rational reconstruction of the interactions between classical foundationalism and science. This reconstruction will be the background for assessing Husserl's version of classical foundationalism.

Aristotle had tried to reconcile deductivism with empiricism by postulating essences in nature which could be known on the basis of experience. However, the ontology of intuitable essences was rejected during the scientific revolution in favour of versions of the corpuscular philosophy.[9] Both Galilei and Descartes argued that the hypothesis of specific essences (for each natural kind) had produced purely verbal knowledge and circular explanations. In order to avoid these evils, a clear distinction had to be made between the *explananda* of science given in experience, and the corpuscular micro-mechanisms that could provide the *explanans* of specific phenomena such as colour or heat in terms of the laws of mechanics. Descartes argued in the second *Meditation* that if the senses cannot reveal to us what matter (e.g. a piece of wax) really is, the first principles of physics cannot be empirical. He concluded that they must be "innate in the mind," like the principles of mathematics.

This rationalist solution to the problem of the first principles took the form of a criterion of truth, according to which clear and distinct intuitions must be true, and a validation of this criterion by the proof that there is an infinite God who does not deceive us with regard to intuitions which we are psychologically unable to reject because of their clarity and distinctness.[10] In his *lettre-préface* to the French edition of *Principia Philosophiae*, Descartes observed that with regard to the real task of philosophers, that is, to find "the first causes and the true principles from which one can deduce the reasons for everything which one is able to know," Plato and Aristotle failed to discover secure principles, whereas he himself succeeded.[11] In the body of that book, he proposed as the fundamental first principles of physics that matter is nothing but extension and that the quantity of motion in the universe is constant.

This rationalist solution to the problem of the first principles soon got into serious trouble. As Arnauld argued in the fourth series of objections to *Meditationes*, Descartes's validation of the criterion of truth by means of a proof of a veracious God begs the question: we must already have validated the criterion in order to accept as true the premises of this proof. Moreover, the principle that matter is nothing but extension implied that there cannot be a vacuum and that, given Descartes's pressure model of light, the propagation of light must be instantaneous. These deductive consequences of the Cartesian principles were empirically refuted in the course of the seventeenth century, the first by Pascal around 1650 and the second by Rømer in 1675. From a philosophical point of view, the refutation by Pascal was especially important, because by dispelling empirically the *ad hoc* explanations which the Cartesians had provided for the barometer experiments, Pascal showed that a testable implication deduced from Descartes's most fundamental principle of physics *alone* had been empirically refuted. This fact demonstrated that the truth of the Cartesian principles could not be guaranteed by Descartes's rationalist criterion of truth, hence, that this criterion is not a viable solution to the epistemological problem of the first principles.[12]

It is no wonder, then, that from 1650 onwards, most progressive scientist-philosophers became empiricists. But their empiricism could not be of the Aristotelian blend, because intuitable essences had been rejected. Modern empiricism had to admit that the basic statements of the sciences are singular statements about observables justified by observation and experiment. By providing these answers to questions 1) and 2) above, modern empiricists exploded the framework of classical foundationalism and therefore could no longer give the deductivist answer to question 3). Newton claimed in his *General Scholium* added in 1713 to Book III of *Principia* that he did not "frame hypotheses" and that he "deduced" his laws from the phenomena. What he meant by "deduction" in this context, however, is inductive derivation, for he continued by saying that in experimental philosophy "particular propositions are inferred from the phenomena, and afterwards rendered general by induction."[13] This Newtonian concept of induction was not Aristotle's notion according to which the mind absorbs general essences. Within the framework of modern empiricism, induction means nothing but a generalization based upon a finite number of singular statements of observation.

According to present-day definitions of 'foundationalism,' Newton's official philosophy of science was foundationalist. But seen from the perspective of classical foundationalism, modern empiricism had to give up the main asset of foundationalism: that it showed how theoretical science can be *knowledge* in the sense of justified (guaranteed) true opinion. Induction in the modern sense cannot guarantee that accepted scientific theories are true. In order to develop a viable alternative to classical foundationalism, then, modern empiricists would have had to provide a new notion of scientific knowledge, a notion that violates the common-sense intuition that an opinion cannot qualify as *knowledge* unless it is true. In Wittgensteinean terms, the empiricist should admit that the 'knowledge' is not homogeneous but a family-resemblance concept. In some contexts, the proposition that X knows that p implies that p is true. But in the context of 'scientific knowledge,' speaking of theoretical knowledge cannot entail that the relevant theories are true.

My meta-epistemology implies that modern empiricism gained superiority over classical foundationalism as a philosophy of science during the second half of the seventeenth century, if measured by the criterion of empirical adequacy. But empirical superiority is not the only criterion for theory choice on this meta-level. Consistency and coherence are important criteria as well. The question which we should answer, then, is whether modern empiricists developed a new coherent theory of scientific knowledge, which showed how induction on the basis of singular observation statements could justify claims to *knowledge*. I shall argue briefly that the British empiricists did not succeed in constructing such a viable alternative to classical foundationalism.

4. AN EMPIRICIST ALTERNATIVE TO CLASSICAL FOUNDATIONALISM? FROM NEWTON TO HUME

Newton's fourth "rule of reasoning in philosophy," added in the third edition of *Principia*, clearly falls short of providing a viable alternative to classical foundationalism. It reads:

> In experimental philosophy we are to look upon propositions inferred by general induction from phenomena as accurately or very nearly true, notwithstanding any contrary hypotheses that may be imagined, till such time as other phenomena occur, by which they may either be made more accurate, or liable to exceptions.[14]

Although Newton was aware of the fact that inductions never guarantee the truth of scientific theories, he still paid lip service to the traditional definition of knowledge as proven true opinion by saying that we should consider the propositions inferred by general induction from phenomena as "accurately true" or "very nearly true."[15] Quite often, Newton uses the term "hypothesis" in the sense of speculations not based upon inductions, such as the Cartesian theory of planetary motion by vortices. This is how we should understand his celebrated claim in *General Scholium* that "I frame no hypotheses." But he failed to say clearly that even propositions inferred by induction remain hypotheses forever and he did not develop a notion of knowledge that incorporates this insight.[16]

The same holds true for Locke and Hume. Although Locke acknowledged the use of hypotheses in natural philosophy, which "often direct us to new discoveries," he retained the old Aristotelian definition of knowledge that inspired the research programme of classical foundationalism.[17] Knowledge, says Locke, implies that "the Mind is possessed of Truth."[18] As we can guarantee truth only by intuition or demonstration, Locke concluded that "whatever comes short of one of these, with what assurance soever embraced, is but Faith, or Opinion, but not Knowledge, at least in all general Truths."[19] It follows that we can have no knowledge of laws of nature or of scientific theories, because we can neither intuit the "real essences" of corpuscular mechanisms, nor deduce propositions about them from perceptual judgments. As Locke says, "we are not capable of scientifical Knowledge; nor shall we ever be able to discover general, instructive, unquestionable Truths concerning them," that is, concerning these corpuscular mechanisms.[20] Clearly, Locke's attempt to force modern empiricism into the straitjacket of classical foundationalism gave rise to scepticism concerning the possibility of scientific knowledge.[21]

Hume did not develop a coherent epistemological alternative to classical foundationalism either. Admittedly, his analysis of induction showed that modern empiricism excludes classical foundationalism. But instead of proposing an alternative account of how scientific theories and predictions might be justified, Hume set out to *explain* by what mental mechanisms human beings become convinced of empirical generalizations and predictions. As these mechanisms allegedly are based upon habit or custom only, and not on reason, Hume implicitly claimed that scientific knowledge cannot be justified at all: the only thing we can do is to explain causally its genesis. In other words, Hume's 'naturalism' does not compensate for his 'scepticism' as far as normative epistemology is concerned.

We might say that Hume is a 'deductivist' in the sense that he never proposed a normative theory of knowledge that might replace classical foundationalism.[22] Even though he implicitly assumed that the inductive method is a valid one, how are we to explain otherwise that he wanted to "introduce the experimental method of reasoning into moral subjects?"[23] His exclusive use of this empirical method prevented him from developing a normative theory of induction: the experimental method applied to human nature could only lead to factual psychological generalizations. Furthermore, there are many passages in Hume's *Treatise* that betray the grip of classical foundationalism on his mind. For example, following the tradition, Hume says that "knowledge" can be obtained only either by intuition or by demonstration, and concludes that no other disciplines than mathematics can constitute a "science."[24]

The situation in epistemology towards the end of the eighteenth century may be summed up as follows. The Cartesian variety of classical foundationalism had been refuted by empirical science, and most contemporaries were convinced of the scientific credentials of Newtonian mechanics. Yet Hume's example had demonstrated that Newton's empiricism could not be a serious epistemological rival for classical foundationalism: it could not show in what sense empirical theories such as Newton's mechanics are *justified* or constitute *knowledge*. No wonder, then, that another great admirer of Newton, Immanuel Kant, attempted to develop a new, non-Cartesian variety of classical foundationalism.

5. KANT, DARWIN, NEO-KANTIANISM

Kant was a classical foundationalist who accepted the Aristotelian requirement that the first principles of the sciences be general and necessarily true. In the absence of intuitable essences, this meant that these principles must be known *a priori*, independently of experience.[25] Kant also assumed, in contrast with Locke and Hume, that Newtonian mechanics is in part a real science, and he tried to demonstrate that the 'pure' part of Newtonian mechanics is based upon synthetic *a priori* principles.[26]

Having concluded, like Descartes, that the first principles of physics cannot be empirical, Kant had to answer question 2) above, how can we know that these principles are true. Kant invented the baroque and complex theory of the first *Critique* in order to explain the 'possibility' of synthetic *a priori* propositions, the details of which need not detain us here. He argued that we are able to know *a priori* the principles of mathematical physics because they inhere in our subjective epistemic mechanism, and that these principles contain information on the physical world (are 'synthetic') because the physical world is partly constituted by this very same subjective mechanism. The theory had many corollaries that Kant found pleasing. It could, for instance, provide an even-handed solution to the problem of the incompatibility of science and religion that had haunted Western thought from the Renaissance onwards.

According to the canonical picture of the history of philosophy, Kant's 'transcendental' theory of knowledge had been refuted by physics at the beginning of the twentieth century, and logical positivists such as Schlick, Carnap, and Reichenbach were the first to see that general relativity and quantum mechanics were incompatible with Kant's idea that Euclidean geometry and the principle of deterministic causality are necessarily true. However, nineteenth-century science had already raised another embarrassing problem for Kantianism. I suggest that this problem explains in part both the specific nature of neo-Kantian speculations in the second half of that century and Husserl's theory of science published in *Ideen* I of 1913. The problem is related to a technical issue in Kantianism that had preoccupied Kantians: the relation between the empirical ego and the transcendental ego.

Kant assumed that all experience must be constituted partly by *a priori* epistemic mechanisms. He had to conclude that apart from our empirical mental life there must be a 'transcendental' mental life or ego, inaccessible to experience because it is an *a priori* constituting condition of all experience. This transcendental ego is the seat of the epistemic structures that constitute experience and yield the *a priori* principles of physics. Since these principles are necessarily true, the transcendental ego cannot change structurally over time and all transcendental egos throughout history must be structurally the same. Each individual allegedly constitutes his or her experience transcendentally. Therefore, each empirical human being must somehow contain an individual transcendental ego. Given Kant's Christian background, these implications did not seem disturbing to him. Could the transcendental ego not be an updated version of the immortal soul, and had God not created human souls for eternity?

However, the assumption of an immutable transcendental structure in human beings became deeply problematical after Darwin. The human mind has evolved from earlier cognitive structures. Consequently, it cannot contain anything immutable. Furthermore, as a thoroughly contingent product of evolution, the structure of the human mind cannot be considered as an explanation of the synthetic *a priori*, because according to Kant's classical foundationalism, synthetic *a priori* judgments are necessarily true. Darwin's theory of evolution raised a dilemma for the followers of Kant. Either Kant's transcendental mechanisms should be downgraded by interpreting them as contingent psychological structures. This option of psychologism had already been chosen by Fries (1773–1843). Or the transcendental structures should be upgraded by promoting them into a realm of real immutability, the realm of Platonic forms. The Platonist alternative became popular among the neo-Kantians. Philosophers of the Marburg school such as Paul Natorp (1854–1924) attempted to establish affinities between Plato and Kant.[27] And the school of Baden, represented by Windelband (1848–1915) and Rickert (1863–1936), postulated an objective and immutable realm of values, apart from the spatio-temporal realm.

It was only this second, Platonist option that could rescue the research programme of classical foundationalism. For this research programme requires that there are some general and synthetic basic propositions that are necessarily true. It was thought that such synthetic *a priori* truths, it was thought, must be concerned with immutable structures or essences, which can subsist only in a Platonic realm. We will come across this same idea in Husserl's foundationalism, as it was developed from his *Prolegomena* (1900) to *Ideen* I (1913).

6. HUSSERL'S FOUNDATIONALISM FROM *PROLEGOMENA* TO *IDEEN* I

In chapter eleven of the first volume (*Prolegomena*) of *Logische Untersuchungen*, Husserl states his foundationalist theory of science (*Wissenschaft*). The theory purports to provide an answer to the question "what makes a science (*Wissenschaft*) into a science," where the extension of the word "*Wissenschaft*" includes not only the natural and human sciences but also mathematics.[28] To this end, Husserl distinguishes between science as an 'anthropological' unity of mental acts or dispositions, and science as an 'ideal' or 'objective' unity. Within this 'objective' unity we may further distinguish two correlative domains that are each unified in its own way: the domain of objects with which an individual science is concerned, and the domain of interrelated truths of which this science consists.[29] According to Husserl, "what makes a science into a science" is a specific *form of interrelation* between the truths of which a scientific discipline consists.

This form is what Husserl calls the "unity of a systematically complete theory."[30] He defines a "complete theory" as an axiomatic deductive system the axioms of which are "fundamental laws" (*Grundgesetze*). A law is "fundamental" if and only if "essentially (that is, "in itself" and not merely subjectively or anthropologically) it cannot be further grounded."[31] Husserl's view of scientific knowledge is in complete agreement with Aristotle's classical foundationalism. Like Aristotle, Husserl defines scientific knowledge as true opinion founded upon a proof, the premises of which are explanatory of the conclusion. Premises are explanatory if they subsume what is stated in the conclusion under a law. If the conclusions are themselves laws, they must be explained by deduction from more elementary laws.[32]

The dangers of a regression or a circle in the chain of deductions are avoided by the assumption that there must be 'fundamental' laws: laws that, *essentially*, cannot be further grounded.[33] Like Aristotle's foundationalism, Husserl's classical foundationalism raises the problem of the first principles. What are the fundamental laws or first principles, and how can they be justified?

In *Logische Untersuchungen*, Husserl provides a clear solution to the problem of the first principles in mathematics (and logic). The first principles or 'fundamental laws' of a mathematical theory are true propositions, the truth of which is directly and exclusively determined by the concepts of which they consist. These principles are justified immediately by an intuition of the essences that allegedly correspond to the fundamental concepts of the theory.[34] To provide this intuitive justification of the principles of mathematics is the task of the philosopher, whereas the mathematician merely constructs deductive systems without being concerned with the justification of fundamental laws. Around 1900, this may have been an acceptable philosophy of mathematics.[35] But Husserl's classical foundationalism runs into difficulties with regard to the empirical sciences. As Husserl himself saw, the empirical sciences risk not qualifying as "scientific knowledge" in his strict sense for the following two reasons.

First, there are many empirical disciplines that are not unified by a theory, let alone a "systematically complete theory." Husserl mentions fields such as geography, astronomy, natural history, and anatomy. Here, the principle of unity that interrelates the truths belonging to such a discipline is "accidental" (*außerwesentlich*): it is not a deductive and explanatory relation between the truths belonging to that science, but, for example, a relation between the objects studied by the discipline. In geography the relevant objects and phenomena are unified by their part–whole relation to the Earth, and in natural history they are unified by the relation of membership of the empirical set of living beings. Yet Husserl concludes his discussion of these matters by stressing again that what is really 'scientific' in these disciplines is their theoretical content.[36]

The second problem with the empirical sciences arises even if they are unified by theories. According to Husserl, a theory in the strict sense is an axiomatic–deductive structure the axioms of which are fundamental laws. But in the empirical sciences, there are no laws of which we know that, 'essentially'; they cannot be grounded by more elementary laws. Even worse, we can never know that empirical theories are true. Husserl concludes that "in the empirical sciences, all theory is mere presumed theory."[37] How, then, is Husserl's classical foundationalism able to account for empirical science?

In *Prolegomena*, Husserl seeks the solution to this traditional problem of classical foundationalism in the theory of probability. He claims that although empirical theories cannot be based upon intuitively *certain* fundamental laws, as is the case in the strictly 'nomological' sciences, they are based upon intuitively *probable* (*einsichtig wahrscheinliche*) fundamental laws, and that, consequently, the empirical theories are themselves intuitively probable.[38] But this solution is confronted by two problems. Husserl does not explain how we may know that empirical laws are 'fundamental' in the defined sense, that is, *essentially* unable to be grounded by deeper laws. Furthermore, Husserl's account of how we may be able to establish *by intuition* the probability of the axioms of an empirical theory such as, for example, classical mechanics, is sketchy at best. Husserl claims that at each historical stage of scientific development the set of available data will determine that *precisely one* theory is the 'only correct' theory. If the data set changes over time, this uniquely selected theory may be a new one, but at each stage there is only one probabilistically justified theory.[39]

It is noteworthy that in the second edition of *Prolegomena*, Husserl changed the term "probabilistic consideration" (*Wahrscheinlichkeitserwägung*) into "empirical consideration" (*empirische Erwägung*).[40] Did he grow dissatisfied with his attempt to save classical foundationalism by a theory of probability? It would be interesting to trace Husserl's ruminations on this topic in the manuscripts which he wrote between 1900 and 1913, and in particular the course on epistemology of 1902–1903.[41] In any case, the probabilistic solution gave way to a very different account in *Ideen* I of 1913, an account that comes close to the neo-Kantian views discussed above.

In the first chapter of *Ideen* I, Husserl develops the notion of a regional ontology. He claims that each factual item has its essential structure. Essences form hierarchies conforming to the so-called law of the inverse variation of intension and extension. The highest genus of such a pyramid of essences is a 'material category' (in opposition to *formal* categories), for example 'physical nature' or 'mental phenomenon.' Since each concrete entity falls under many highest genera – for instance, an individual bird will fall under the categories of life, colour, three-dimensional object, etc., there is for each concrete entity a complex structure of interconnected highest genera or material categories. Husserl calls such a structure a "region of being."[42] He claims that we might investigate these regions by *Wesensschau* or eidetic intuition, and that the resulting truths are both synthetic and *a priori*.[43] The class of truths concerning a particular region constitutes a regional ontology.

In *Ideen* I, Husserl solves the problem of the first principles of the empirical sciences by means of this notion of a regional ontology. He claims that each regional ontology has a number of fundamental truths (*Grundwahrheiten*), its axioms, that apply apodictically and with unconditional necessity to each of the individual objects belonging to this region.[44] Since these truths are directly justified by *Wesensschau*, they may serve as first principles to the factual or empirical sciences: "each factual science (empirical science) has essential theoretical foundations in eidetic ontologies."[45] Indeed, to the extent that an empirical discipline approaches the ideal of a theoretical or 'nomological' science, it will be based upon such eidetic foundations, which are necessarily true.[46]

From this review of classical foundationalism, we may conclude that with Husserl the development of this research programme has completed a full circle. Initially, Plato and Aristotle solved the problem of the first principles by postulating essences. During the scientific revolution, this solution was rejected. After Kant and Darwin, the hypothesis of essences became popular again as a means for securing foundations for the sciences, foundations that are both general and necessarily true. Husserl's foundationalism fits in with this fashion of Aristotelian, or rather Platonist, classical foundationalism.

7. HEIDEGGER AND THE DEGENERATION OF CLASSICAL FOUNDATIONALISM

What was the rationale for Husserl's essentialism? What was its appeal to Husserl and his followers in the Göttingen and Munich circles of phenomenologists? Did essentialist classical foundationalism qualify as a justified epistemology of mathematics and science at the time Husserl proposed it, around 1913? Finally, could classical foundationalism survive the scientific revolutions in physics of the first quarter of the twentieth century?

Husserl's essentialism originated, like Plato's theory of *eide*, in the philosophy of mathematics, but the branches of mathematics that inspired these two philosophers were different. Plato's initial question was concerned with geometry. If geometry is knowledge of *perfect* spatial forms, such as perfectly round circles or perfectly straight lines, its objects cannot be located in the imperfect empirical world. Hence there must be another, spiritual world, in which these perfect objects reside, objects that serve as paradigms for empirical technologies aiming at constructing spatial artifacts. Plato then generalized the doctrine of perfect paradigms to all empirical phenomena (even mud has its Form). In Husserl's case, essentialism was inspired by number theory. Having rejected in *Logische Untersuchungen* the Lockean theory of the 1891 *Philosophie der Arithmetik*, according to which number concepts are abstracted from introspected individual acts of 'collecting,' Husserl argued in 1900–1901 that numbers are essences of sets, although while writing the book, he changed his mind concerning the question as to what *precisely* numbers are essences of.[47] The difference in origin between Plato's and Husserl's theories explains an important difference between the theories themselves. Whereas according to Plato the Forms are more perfect than their instances, Husserl rejected this tenet of Platonism. Obviously, it would not make sense to say that the number 5 is 'more perfectly five' than an individual set of five members.[48]

There can be no doubt that around 1900, the philosophical idea that number theory is true of an objective domain of Platonic entities, sets-as-types, was justified in the relative sense explained in the introduction to this paper. Psychologism *à la* John Stuart Mill was the only serious competitor on the market. But Mill's empiricist view that logic and mathematics consist of empirical regularities governing mental acts had been refuted by Husserl (*Prolegomena*) and Frege. Even twentieth-century philosophers of mathematics such as Boolos or Quine accepted the existence of sets-as-types.[49] Wittgenstein's radical alternative in the philosophy of logic and mathematics, according to which these disciplines are not *about* a domain of entities at all since they resemble parts of grammar, was not yet available.

Much more problematic, however, is Husserl's generalization of his doctrine of essences to all factual entities. We saw that, in 1913, Husserl defended the doctrine that each individual empirical entity falls under a number of essences, and that regional ontologies of these essences provide the first principles of the nomological sciences. We may reconstruct Husserl's reasons for this generalization as follows. In the philosophy of logic, Husserl defended the doctrine that apophantic logic studies a domain of objects and their interrelations: the domain of propositions. According to *Logische Untersuchungen*, propositions are essences or types, the tokens of which are time-bound mental acts of judging, which give meaning to physical signs. This view enabled Husserl to explain how the laws of logic that are concerned with propositions as a-temporal types, could be relevant to our mental life in time. The view implies that mental acts fall under essences, and it was one of the tasks of Investigations 1, 5, and 6 to identify the particular aspects of mental acts of which propositions are the types. However, if some aspects of mental acts have types or essences under which they fall, should not all aspects of all empirical phenomena fall under essences? And should we not assume that there are *a priori* disciplines like logic or number theory that study these 'material' essences? This was precisely Husserl's conclusion in 1913. Ultimately, then, it was Husserl's philosophy of logic and mathematics that inspired his idea of regional ontologies as foundations of the empirical sciences.

Although none of Husserl's pupils was prepared to follow the master on the road to transcendental idealism, they all accepted the doctrine of essences.[50] Whatever the logical force of the arguments which Husserl adduced in favour of essentialism in his Second *Logische Untersuchung*, the doctrine of essences had the advantage of vindicating a foundational role for philosophers at a time in which the emancipation of psychology from philosophy threatened to reduce the latter field to insignificance. Essentialism underpinned in a new way the traditional claim of philosophers that they alone were able to lay secure foundations for the sciences. Unfortunately, however, the essentialist variety of classical foundationalism had a serious disadvantage, which the development of physics would bring to light all too soon. This disadvantage may be formulated as a dilemma. Either the philosopher is able to intuit essences by *Wesensschau*, in which case, the fundamental laws which he or she discovers will be valid essentially and forever, and the foundations of the sciences cannot vary over time, or, if the foundations of the sciences vary over time, *Wesensschau* is an unreliable procedure for securing the foundations of the sciences and the hypothesis that there are essences is implausible. In no case can essentialism cope with the phenomenon of scientific revolutions.[51]

By 1927, when Heidegger published *Sein und Zeit*,[52] philosophers had become aware of the theoretical revolutions in physics that had shaken the intellectual landscape in the first quarter of the twentieth century. It is fascinating to read section three of *Sein und Zeit*, since in that section Heidegger attempts to reconcile the phenomenon of scientific revolutions with Husserl's notion of regional ontologies. How can the philosopher acknowledge on the one hand that the fundamental concepts and principles of scientific disciplines may be changed by empirical scientists, and maintain on the other hand that philosophy is more fundamental than the sciences because it lays their foundations? As I have argued elsewhere, in *Sein und Zeit* Heidegger does not succeed in synthesizing these two irreconcilable tenets. [53]

Elsewhere, Heidegger laconically replies to the questions as to how philosophy might accommodate the phenomenon of scientific revolutions, how a science may have different "fundamental attitudes" (*Grundstellungen*) to its subject matter and how a revision of fundamental concepts expresses a shift in attitude.[54] These empty formulations are symptomatic of the fact that, in Heidegger's time, the research programme of classical foundationalism had reached the stage of its final decline, dragging down in its bankruptcy the notion that philosophy is fundamental to the sciences. Its most promising rival as an epistemology of the empirical sciences is not coherentism, but competitive empiricism, as discussed in the introduction to this essay. Within the framework of competitive empiricism, the problem of the first principles simply does not arise.

Notes

1   See Paul R. Thagard, "Why Astrology Is a Pseudoscience," in *Philosophy of Science: The Central Issues*, ed. Martin Curd and J.A. Cover (New York, London: W.W. Norton & Company, 1998): 31–35.

2   See Robert Audi, *Epistemology* (London and New York: Routledge, 1988); Jonathan Dancy, *Introduction to Contemporary Epistemology* (Oxford: Basil Blackwell, 1985); Charles Landesman, *An Introduction to Epistemology* (Oxford: Blackwell, 1997).

3   The present paper may be considered as a complement to my paper "Transcendental Idealism," in which I trace the origins of Husserl's transcendental idealism. This paper is in *The Cambridge Companion to Husserl*, ed. Barry Smith and David Woodruff Smith (Cambridge: Cambridge University Press, 1995).

4   Dancy, *Introduction to Contemporary Epistemology*, 53.

5   H. Scholz "The Ancient Axiomatic Theory," in *Articles on Aristotle*, vol. 1, *Science*, ed. J. Barnes, M. Schofield, and R. Sorabji (London: Duckworth, 1975). Scholz discusses the application of Aristotle's foundationalism to mathematics and speaks of 'classical axiomatics.'

6   The research programme of classical foundationalism (the 'quest for certainty') is not based upon conceptual confusions, as linguistic philosophers have argued. See D.W. Hamlyn, *The Theory of Knowledge* (London: The MacMillan Press, 1970), 10–15.

7   Hans Albert, *Traktat über kritische Vernunft*, 4th ed. (Tübingen: J.C.B. Mohr, 1980), 13 and *passim*.

8   Apparently, Aristotle considers the finiteness of the human mind as the crux of the difficulty, but this is mistaken. Even an infinite mind, that might be able to go through infinitely many things at a time, cannot establish the truth of the conclusion if there is an infinite regress in the chains of deductions.

9 Although authors such as Descartes and Locke paid lip service to the jargon of essences, they did not think that intuition of specific essences can provide the basic premises of scientific deductions by which particular phenomena are explained. For Descartes, there are but two essences, those of matter and mind, and according to Locke, real essences are unknowable.

10 In fact Descartes did not think that he could deduce explanations of specific phenomena such as light from his first principles, although he amply used the rhetoric of classical foundationalism. The principles rather determine a logical space of admissible physical hypotheses. Hence these hypotheses cannot possess more than "moral" certainty. See René Descartes, *Les principes de la philosophie: Œuvres philosophiques*, ed. F. Alquié (Paris: Garnier Frères, 1973), 4: 204–206. Descartes proposes a theological version of the principle of underdetermination of theories by experience.

11 Descartes, « Lettre-préface de l'édition française des Principes » (« Lettre de l'auteur à celui qui a traduit le livre laquelle peut ici servir de préface »), *Œuvres philosophiques*, 3: 772–773.

12 E.J. Dijksterhuis, *De mechanisering van het wereldbeeld* (Amsterdam: Meulenhoff, 1950), 261–278.

13 Isaac Newton, *Mathematical Principles of Natural Philosophy*, tr. Andrew Motte, rev. Florian Cajori, 2 vols. (Berkeley: University of California Press, 1962), Book III, 547.

14 Newton, *Mathematical Principles of Natural Philosophy*, Book III, 400.

15 That Newton was aware of the fact that inductions do not guarantee the truth of their conclusions is demonstrated by Query 31 to *Opticks*. "And although the arguing from Experiments and Observations by Induction be no Demonstration of general Conclusions; yet it is the best way of arguing which the Nature of Things admits of, and may be looked upon as so much the stronger, by how much the Induction is more general. And if no Exception occur from Phaenomena, the Conclusion may be pronounced generally." *Opticks or A Treatise of the Reflections, Refractions, Inflections, & Colours of Light* (New York: Dover Publications, 1979), 404.

16 Newton's failure at this point is shown by many passages, such as Query 28 to *Opticks*, where Newton writes, having condemned as a "hypothesis" the Cartesian theory of a "dense Fluid" in which the planets and comets move: "Whereas the main Business of natural Philosophy is to argue from Phaenomena without feigning Hypotheses, and to deduce Causes from Effects, till we come to the very first Cause, which certainly is not mechanical." Newton then develops an Argument from Design, and shows no awareness at all that the conclusion that God exists is in fact a hypothesis

that is not better supported than the physical speculations of the Cartesians.

17 John Locke, *An Essay concerning Human Understanding,* ed. Peter H. Nidditch (Oxford: Oxford University Press, 1975), 648.

18 Locke, *An Essay concerning Human Understanding,* 527.

19 Locke, *An Essay concerning Human Understanding,* 536–7.

20 Locke, *An Essay concerning Human Understanding,* 557, 588.

21 Michael Ayers, *Locke: Epistemology and Ontology* (London and New York: Routledge, 1993), 1: 118, explains Locke's "pessimism about science," which contrasts with Bacon's optimism, by saying that "[Locke's] doctrines represent an effective critical assessment of what had been achieved to date rather than a visionary programme for the future." I think that there is a better explanation, based on the fact that Locke implicitly used Aristotelian foundationalism as a criterion for the possibility of knowledge. Locke's theory of perception implies that it is *a priori* impossible to have an intuition of 'real essences,' that is, of the corpuscular mechanisms that cause ideas in us. But without an intuitive knowledge of essences, scientific deductions in Aristotle's sense are impossible. Hence, natural science can never amount to 'knowledge,' irrespective of the new experiments which future scientists might conduct. There is no reason, then, to suppose with Ayers that Locke merely provided a "critical assessment of what had been achieved to date" in the sciences, and there is no support in the text of the *Essay* for this assumption.

22 In the secondary literature on Hume, there is an ongoing debate on whether Hume was a deductivist. See D.C. Stove, *Probability and Hume's Inductive Scepticism* (Oxford: Oxford University Press, 1973). For a synopsis of this debate, see Peter Millican, "Induction," in *Encyclopedia of Empiricism*, Don Garrett and Edward Barbanell (London and Chicago: Fitzroy Dearborn Publishers, 1997).

23 David Hume, *A Treatise of Human Nature*, ed. L.A. Selby-Bigge, rev. P.H. Nidditch (Oxford: Oxford University Press, 1978), title page. Following Newton, Hume also claimed to reject "hypotheses": "And tho' we must endeavour to render all our principles as universal as possible, by tracing up our experiments to the utmost, and explaining all effects from the simplest and fewest causes, 'tis still certain we cannot go beyond experience; and any hypothesis, that pretends to discover the ultimate original qualities of human nature, ought at first to be rejected as presumptuous and chimerical." xvii.

24 Hume, *A Treatise of Human Nature*, 69–73.

25 Immanuel Kant, "Metaphysische Anfangsgründe der Naturwissenschaft," in *Werke*, ed. Wilhelm Weischedel, Bd. 8 (Darmstadt: Wissenschaftliche Buchgesellschaft, 1968), A vi: "Eine rationale Naturlehre verdient also den Namen einer Naturwissenschaft nur alsdenn, wenn die Naturgesetze, die in ihr zum Grunde legen, a priori erkannt werden, und nicht bloße Erfahrungsgesetze sind."

26 Kant, *"Metaphysische Anfangsgründe der Naturwissenschaft,"* A vi.

27 See Paul Natorp, *Platons Ideenlehre: Eine Einführung in den Idealismus* (Leipzig: Dürr and Meiner, 1903).

28 "was Wissenschaft zur Wissenschaft macht," Edmund Husserl, *Logische Untersuchungen*, 2 vols. (Halle: Max Niemeyer Verlag, Erste Auflage, 1900 and 1901, Zweite Auflage, 1913 and 1921), *Logische Untersuchungen, Erste Auflage*, I, 228.

29 Husserl, *Logische Untersuchungen, Erste Auflage*, I (*Prolegomena*), § 62.

30 Husserl, *Logische Untersuchungen, Erste Auflage*, I, § 63, 232.

31 Husserl, *Logische Untersuchungen. Erste Auflage*, I, § 63, 232: "Die Begründung von generellen Gesetzen führt notwendig auf gewisse, ihrem Wesen nach (also "an sich" und nicht bloß subjektiv oder anthropologisch) nicht mehr begründbare Gesetze. Sie heißen Grundgesetze."

32 Husserl, *Logische Untersuchungen, Erste Auflage*, I, § 63, 231: "Wissenschaftliche Erkenntnis ist als solche *Erkenntnis aus dem Grunde*. Den Grund von etwas erkennen, heißt die Notwendigkeit davon, daß es sich so und so verhält, einsehen. Die Notwendigkeit als objektives Prädikat einer Wahrheit ... bedeutet soviel wie gesetzliche Gültigkeit des bezüglichen Sachverhaltes."

33 Husserl, *Logische Untersuchungen, Erste Auflage*, I, Anhang ad §§ 25–26, 84–85: "Beruht aber jede Begründung auf Prinzipien, denen gemäß sie verläuft, und kann ihre höchste Rechtfertigung nur durch Rekurs auf diese Prinzipien vollzogen werden, dann führte es entweder auf einen Zirkel oder einen unendlichen *Regreß*, wenn die Begründungsprinzipien selbst immer wieder der Begründung bedürfen ... Also ist es evident, daß die Forderung einer prinzipiellen Rechtfertigung für jede mittelbare Erkenntnis nur dann einen möglichen Sinn haben kann, wenn wir fähig sind, gewisse letzte Prinzipien einsichtig und unmittelbar zu erkennen, auf welchen alle Begründung im letzten Grund beruht." Husserl is discussing the principles of logic, although he uses this same foundationalist argument throughout.

34 For this interpretation see Herman Philipse, *De fundering van de logica in Husserls "Logische Untersuchungen,"* (Leiden: Labor Vincit, 1983), §§ 9–10. Husserl argues in *Logische Untersuchungen* that the objects of logic and

mathematics are essences with which we may be acquainted by what he calls *"Wesensschau"* or "Ideation."

35 Russell proposed a similar theory in 1903. Bertrand Russell, *The Principles of Mathematics* (London: Allen & Unwin, 1937), xv. However, Husserl's doctrine of a division of labour between mathematicians and philosophers was superseded when mathematicians started to develop meta-mathematics in the beginning of the twentieth century. Modern axiomatics drop Aristotle's evidence postulate in favour of a rigorously formulable postulate of consistency. With regard to the requirement that axioms must be necessarily true, Brouwer and Weyl sided with Aristotle and Husserl, whereas Hilbert rejected the necessity requirement. See Scholz, "The Ancient Axiomatic Theory," §§ 9 and 14.

36 Husserl, *Logische Untersuchungen, Erste Auflage*, I, § 64.

37 Husserl, *Logische Untersuchungen, Erste Auflage*, I, § 72, 255: "Alle Theorie in den Erfahrungswissenschaften is bloß supponierte Theorie."

38 Husserl, *Logische Untersuchungen, Erste Auflage*, I, § 72, 255, where Husserl says about theories in the empirical sciences: "Sie gibt nicht Erklärung aus einsichtig gewissen, sondern nur aus *einsichtig wahrscheinlichen* Grundgesetzen. So *sind die Theorien* selbst nur von einsichtiger Wahrscheinlichkeit... ."

39 Husserl, *Logische Untersuchungen, Erste Auflage*, I, § 72, 256: "Wir erheben den Anspruch, daß es jeweils nur ein berechtigtes Verhalten in der Wertschätzung der erklärenden Gesetze und in der Bestimmung der wirklichen Tatsachen gebe, und zwar für jede erreichte Stufe der Wissenschaf" and "Im Bereiche früherer Erfahrung war die frühere, im Bereiche der erweiterten Erfahrung ist die neu zu begründende Theorie die 'einzig richtige,' sie ist die einzige durch korrekte Wahrscheinlichkeitserwägung zu rechtfertigende."

40 Compare the last quoted sentence in the previous note with Husserl, *Logische Untersuchungen, Zweite Auflage*, I, 257: "sie ist die einzige durch korrekte empirische Erwägung zu rechtfertigende."

41 This course will be published soon in *Husserliana*, Materialband III.

42 Husserl, *Ideen zu einer reinen Phänomenologie und phänomenologischen Philosophie, Erste Buch. Allgemeine inführung in die reine Phänomenologie*, ed. Karl Schuhmann, *Husserliana*, vol. III/1 (den Haag: Martinus Nijhoff, 1976), § 16, 36: "Region ist nichts anderes als die gesamte zu einem Konkretum gehörige oberste Gattungseinheit, also die wesenseinheitliche Verknüpfung der obersten Gattungen, die den niedersten Differenzen innerhalb des Konkretums zugehören." See also § 9.

43 Husserl, *Ideen*, 36. Husserl defines "analytical" necessities as formal laws, in which the terms may be varied freely *salva veritate*.

44 Husserl, *Ideen*, 37.

45 Husserl, *Ideen*, § 9, 23: "Jede Tatsachenwissenschaft (*Erfahrungswissenschaft*) hat wesentliche theoretische Fundamente in eidetischen Ontologien." (Husserl's emphasis).

46 Husserl, *Ideen*, 24.

47 While writing *Prolegomena*, Husserl held that numbers are essences, *ideale Spezies*, of mental acts of counting. But in chapter six of the Sixth Logical Investigation, he defends the doctrine that numbers are essences of the categorial forms of sets, which are *intentional correlates* of acts of counting, and this remains Husserl's conviction afterwards. One might trace Husserl's change of mind by comparing the second edition of *Prolegomena* (1913) with the first (1900). For instance, on pp. 170–171 of *Logische Untersuchungen, Erste Auflage*, I, Husserl writes: "Die Zahl Fünf ist … die ideale Spezies, die in gewissen Zahlungsakten ihre konkreten Einzelfälle hat." In the second edition this sentence is expanded as follows: "Die Zahl Fünf ist … die ideale *Spezies* einer Form, die in gewissen Zählungsakten auf Seiten des in ihnen Objektiven, des konstituierten Kollektivum, ihre konkreten *Einzelfälle* hat."

48 See Husserl, *Logische Untersuchungen, Erste Auflage*, II, *Erste Untersuchung*, § 32.

49 According to Husserl, a token set is not a Platonic object but an empirical entity with categorial aspects, constituted by a hierarchy of mental acts (perceptions or imaginations of the member-entities, upon which is founded a categorial act of collecting these entities into a set). For this theory, see the Sixth Logical Investigation, §§ 40ff.

50 For a reconstruction of Husserl's road to transcendental idealism, see Philipse, "Transcendental Idealism."

51 For a Wittgensteinean critique of essences, see Philipse, *Heidegger's Philosophy of Being: A Critical Interpretation* (Princeton: Princeton University Press, 1998), 335–341.

52 Martin Heidegger, *Sein und Zeit* (Tubingen: Max Niemeyer Verlag, 1967).

53 Philipse, *Heidegger's Philosophy of Being*, 36–39.

54 Philipse, *Heidegger's Philosophy of Being*, 404 n.148.

CHAPTER TWO

Sonja Rinofner-Kreidl

WHAT IS WRONG WITH NATURALIZING EPISTEMOLOGY? A PHENOMENOLOGIST'S REPLY

---

1. INTRODUCTION

In recent philosophical disputes the odds are on the naturalist's side. Despite regular complaints about the ambiguity of the term 'naturalism,' there exists a tacit agreement among scientists and a growing group of philosophers that defending some version of naturalism is the only tenable position. I shall not discuss the prospects of naturalism in general; rather, I shall criticize the idea of naturalizing epistemology as considered in the light of transcendental phenomenology. My examination of Quine's and Husserl's views focuses on the problems of circularity and foundationalism and concludes with a sketch of how naturalized epistemology and phenomenology handle the problem of scepticism. I do not aim at presenting either a detailed critique of naturalized epistemology or a detailed interpretation of transcendental phenomenology. I am primarily interested in the following questions. Could naturalists and anti-naturalists reach an agreement about how to formulate the problem of naturalizing epistemology, and what procedure to follow in order to bring it into focus? Suppose this metatheoretical conciliatory attempt fails, are we then urged to concede that there is no rational and non-arbitrary way of settling the matter? Do we then have to plead for begging the question?

A useful way of specifying different types of naturalism is by distinguishing ontological (metaphysical), semantical (analytical), and methodological naturalism. Ontological naturalism asserts that only the natural sciences (or one particular natural science) are entitled to single out a privileged class of entities, so-called natural entities. Semantical naturalism contends that no philosophical theory is acceptable as long as its key concepts (e.g. intentionality) have not been shown to be analysable in terms of naturalistically-approved concepts. Methodological naturalism (primarily identified with Quine's epistemology) asserts that no philosophical theory of knowledge can pretend to achieve any special, nonempirical knowledge, and that any philosophical interpretation of scientific theories, if needed at all, must occur within the domain of the empirical sciences. Where does Husserl's anti-naturalistic transcendental phenomenology stand with respect to this tripartite scheme of naturalisms? Transcendental phenomenology surely criticizes ontological naturalism. Nonetheless, it does not support any version of ontological anti-naturalism because it is based on a primacy of epistemology opposed to ontology. Transcendental phenomenology equally criticizes the project of semantical naturalism from the point of view of a descriptive analysis. In my view transcendental phenomenology defends a special version of methodological anti-naturalism. In addition to the previous types of anti-naturalism, it is common to distinguish radical and moderate forms of naturalism or strong(er) and weak(er) programmes of naturalizing.[1] Ontological naturalism, for instance, is a stronger position than methodological naturalism. (Of course, we may discover hidden ontological commitments in certain realizations of methodological naturalism.) What is meant exactly by 'radical' and 'moderate,' or 'strong' and 'weak,' depends on our classification of the basic types of naturalism.

Methodological naturalism and methodological anti-naturalism both seem to be attractive approaches. Their common ground lies, *prima facie*, in cautiously abstaining from absolute, dogmatic positions. I take it that Quine's relativization or elimination of conceptual distinctions which were of central importance to traditional epistemologies, e.g. formal/material, empirical/*a priori*, analytic/synthetic, descriptive/normative, is a "radical naturalism." Its radical nature lies in the fact that a Quinean critique of traditional concepts includes a thoroughgoing shift of those problems which are deemed to be philosophically or theoretically worth investigating. If it is true that Quine's naturalizing epistemology project unhinges the fundamental conceptual distinctions of traditional epistemology, then naturalists and anti-naturalists argue on the basis of radically divergent ideas of knowledge. This is by no means surprising. What essentially is at stake when we discuss the idea of naturalizing epistemology is the concept of epistemology.[2] Yet, this being the case, it becomes unclear how the problem could be formulated and solved in a way that does justice to both sides.

Quine considers the opposition between naturalism and traditional philosophy to be roughly tantamount to the opposition between a theory primarily labeling physical objects and a sense-data theory. This view is obviously not appropriate for understanding the conflict between a naturalistic research programme and transcendental phenomenology. It might be possible to understand these views referring to the opposition between positive science and Cartesian foundationalism. But this would be misguided, as Quine himself admits, stating that he would accept Cartesian dualism as a scientific hypothesis only if there were some (indirect) explanation–value in positing an essentially non-material spirit.[3] Naturalism should not be identified with materialism. In sum: Cartesian-style foundationalistic projects fail and the same holds for sensualistic foundationalism. Transcendental phenomenology defends neither sensualism nor a renewed Cartesianism. Yet how can we explain that phenomenology is to a large extent ignored as a serious alternative to naturalized epistemology? It seems partly because transcendental phenomenology is related to the idea of justificational foundationalism.[4] Nevertheless, the strategy of opposing naturalism and foundationalism (which makes Quine's challenge so attractive – "Why not settle for psychology?") does not work with regard to transcendental phenomenology.

2. THE CIRCULARITY DEBATE

A phenomenological anti-naturalist would argue as follows. If we assume methodological naturalism, then we only investigate psychological processes of achieving, enlarging, corroborating, doubting, and refuting knowledge. However, such an empirical investigation necessarily presupposes that we can have valid knowledge. Explaining what it means to have valid knowledge is the appropriate task of epistemology. Husserl's anti-naturalism demands that different types of questions be treated as different. Therefore, in each case we can discern whether we have different answers to the same question or different questions. Pointing to empirical investigations in the field of psychology, physiology, biology, etc. does not help if we want to know how knowledge *per se* is possible. Empirical investigations implicitly presuppose the reality of both the things investigated and the investigators. However critical scientific methods may be, and whatever pains-taking analyses they may engender, this tacit reality thesis (which is like a naïve, or hypothetical, realism) remains untouched. Consequently, if we pretend to answer the question of how it is possible to acquire knowledge about real things without committing the fallacy of *petitio principii*, we must neutralize the reality thesis. Ignoring it entails, Husserl says (embracing an Aristotelian figure of argumentation), a *metabasis eis allo genos,* an unnoticed shifting of the object of investigation (or proof).[5] A metabasis marks a deficient, invalid line of reasoning. Trying to answer *quid iuris* questions we have to make sure not to fall prey to a metabasis. This is precisely the purpose of Husserl's much-disputed phenomenological reduction, which is a methodological device for avoiding a metabasis.[6]

A Quinean-style naturalist argues that any epistemology pretending to proceed in a pure, presuppositionless manner, dissociated from empirical scientific research, is doomed to failure.[7] Contrary to the anti-naturalist's intention any such project inevitably includes *one* presupposition. Pure epistemology must assume its own reasonable character. It has to assume that pure conscious content can be analysed without thereby taking notice of real cognitive processes. But presupposing this means to take for granted that knowledge *is* possible because epistemology is taken to be possible, and epistemology, when carried out, is a piece of knowledge. Since traditional epistemology does not realize that it inevitably hinders its own realization precisely by trying to bring it about, it is a self-delusive undertaking. Any aspiring after principal foundations of knowledge must presuppose knowledge to be already well-founded.[8] Otherwise, the epistemological project fails right from the beginning. On the other hand, epistemology seems to be unnecessary if the possibility of knowledge has to be presupposed whenever one tries to prove this very possibility. It is impossible to explain how our judgments could correspond to their objects *without presupposing this correspondence.* We do not even know how we could come to know a putatively missing correspondence otherwise than by having recourse to our ordinary, corrigible knowledge, e.g., to fallible predictions of future occurrences based on empirical hypotheses. This being the case, why should we care for a possibly missing correspondence? There is no need to ask for the conditions of the possibility of knowledge.

The most general distinguishing feature of naturalism is its insistence on "natural phenomena." According to Quine, epistemology is part of an ongoing process of structuring and restructuring knowledge. There is no privileged position in this striving after knowledge and no privileged type of knowledge, no absolutely certain and no *a priori* knowledge. Epistemological questions are not in any sense more fundamental than the rest of our theoretical views. Empirical and philosophical investigations are on a continuum. To embrace the natural character of all our knowledge is to reject the idea of an extra-worldly view that we could use to independently and objectively evaluate the methodical standards, criteria, and results of our knowledge.[9] Proceeding on these lines, "naturalizing epistemology" primarily means to de-mystify and de-transcendentalize epistemology. One remarkable characteristic of Quine's naturalism is that he does not try to weaken or to refute the circularity objection raised against his theory according to the anti-naturalistic argumentation sketched above. Quine explicitly emphasizes that we have to accept the circular structure of all our scientific knowledge which is grounded in the relation existing between observation sentences and scientific theories.

> The stimulation of his sensory receptors is all the evidence anybody has had to go on, ultimately, in arriving at his picture of the world. Why not just see how this construction really proceeds? Why not settle for psychology?

> Such a surrender of the epistemological burden to psychology was previously disallowed as circular reasoning. If the epistemologist's goal is to validate the grounds of empirical science, he defeats his purpose by using psychology or other empirical science in the validation.

> However, such scruples against circularity have little point once we have stopped dreaming of deducing science from observations. If we are to simply understand the link between observation and science, we are well advised to use any available information, including that provided by the very science whose link with observation we are seeking to understand.[10]

These remarks make clear that Quine's naturalism is essentially grounded in a holistic view of how scientific theories are produced and tested. This view is primarily opposed to naive empiricism and sensationalism.[11] If we give up these dogmas the problem of circularity does not vanish, but loses all force. The same holds true for the quarrels between rationalists and empiricists, idealists and realists, which have long encumbered any productive scientific work.

As is often the case in strongly polarized discussions, naturalists tend to overemphasize opposite views. In defending naturalism by minimizing reasonable alternatives one may, for instance, oppose hypothetical–empirical knowledge with mind-independent, eternal truth. One could explain this strategy by referring to Otto Neurath's boat metaphor, which rejects the *tabula rasa* idea and asserts the immanence of knowledge.[12] Quine adopts this metaphor, which states that we either stay in the boat and do empirical research or step outside and drown with our heads full of the rationalistic ideas of an *a priori* 'first philosophy.' But the seemingly hopeless situation of 'first philosophy' is because the boat metaphor begs the question. Why should we take for granted that asking for non-empirical knowledge is, on principle, illegitimate? In my view, the gist of what phenomenology has to say against naturalism is that it rejects the naturalistic strategy of opposing radically divergent positions by tacitly assuming them to cover the whole range of possible approaches (e.g. holistic naturalism vs. naïve empiricism or holistic naturalism vs. extreme rationalism).

3. PHENOMENOLOGICAL FOUNDATIONS

If all our knowledge is, as the naturalist argues, corrigible, then we may ask whether this openness to future experience also applies to the naturalist's exclusion of *a priori* knowledge. What is the epistemic status of the thesis that all knowledge we can achieve is empirical knowledge? Does the naturalist admit this thesis could be corrected by future investigations? In other words: is the naturalist's assertion that there is no *a priori* knowledge empirical or *a priori* knowledge? In order to avoid a contradiction the naturalist will insist that his denial of *a priori* knowledge is provisional and based on empirical knowledge. Hence, the fundamental naturalistic claim that we ought not to look for anything else but empirically testable hypotheses cannot represent, strictly speaking, a general principle. It has to be considered a guiding principle of our scientific research, which falls under the fallibility-condition in the same way as the results of the research itself. Emphasizing the role of experience obviously indicates that hitherto unknown results of future scientific investigations could devaluate even the most general forms of representing objects. Let us take an example in order to understand how the empirical and phenomenological points of view differ. Imagine me sitting in my bureau at Graz University working on Quinean matters. In this situation it may happen that I am suddenly distracted by a mental image of the Hundertwasser Tower I saw the day before in Vienna. Although perceptually aware of my present surroundings (at least in terms of some dim background awareness), I am simultaneously confronted with a fantasy image that I immediately interpret as a memory. Physiological and psychological investigations can explain what, under these particular circumstances, probably has caused the appearance of the fantasy image. But whatever knowledge we achieve by means of such investigations, empirical and phenomenological analyses will always remain clearly distinct. If I decide to take up the latter, I am interested in the general structure of different types of intentions, for instance sensual perceptions, imaginations, rememberings, judgings, and so on. No psychological or brain-physiological investigation could ever offer a functional equivalent to describing the forms of different types of intentionality "from within," i.e. from the perspective of a perceiving, remembering, judging person. The phenomenon implies a subject-relatedness which gets lost when "transferred" to the methodical framework of psychological or brain-physiological investigations. Phenomenology attempts to find some *a priori* truth concerning the general structure of objects, and correspondingly, their modes of appearance.

It is important to see that phenomenology, notwithstanding its aprioristic character, is to some extent open to future experience. Understanding this demands the recognition of the special brand of phenomenological apriorism that rejects the naturalistic strategy of *minimizing reasonable alternatives*. The choices are not limited to naturalized epistemology *à la* Quine, a naive empiricism, and a radical rationalism. Phenomenology advocates widening the concept of "experience" beyond that of sensual experience, thus permitting non-sensual experience referring to ideal objects indirectly situated in space and time *"sekundär lokalisiert,"* i.e. by means of sensual objects to which they are connected.[13] Therefore, phenomenological apriorism is committed to a special type of experience. It is not, on principle, independent of any experience whatever, as Kant considers *a priori* knowledge to be. Transcendental phenomenology differs from the Kantian view of transcendental philosophy in some essential respects. These differences mainly ground the intuitive and descriptive method of phenomenology. As Husserl clearly sees, this method does not allow for establishing any definite and complete system or set of first principles or categories which could be proved necessary for any possible objective experience. Considering the range of problems and concepts transcendental phenomenology intends to analyse, it is not entirely independent of the empirical sciences which deliver the "material" for philosophical analyses. So far as the formal status of the objects analysed is concerned, phenomenology is completely independent of empirical science (or positive science in general). Transcendental phenomenology supports a descriptive apriorism which is akin to empirical research in one important respect: it is work in progress. Phenomenological descriptions do not reach a final state of enclosure in a so-called philosophical system. However, they cannot be reduced to or integrated into empirical or purely formal scientific investigations. Phenomenological judgments do not represent metaphysical knowledge either. A phenomenologist says: if we experience an object x, this object or the experience of this object necessarily shows a certain essential structure (w, y, z, etc.). Otherwise, it would not be an object of this-or-that-kind or an experience of this-or-that-kind. But whatever the phenomenologist may find out concerning the essences of objects and intentional structures he, thereby, does not prove that the essences in question necessarily have to be realized, i.e. that we actually must have an experience of the objects in question.[14]

Eidetic phenomenological judgments do not pretend to give an absolutely certain knowledge. In this view Husserl, if rightly understood, agrees with the naturalist's statement that "there is no infallible knowledge and there is no absolute objectivity."[15] Phenomenology does not deny that all our knowledge is a "natural phenomenon," if this would mean to deny the reality of those acts and actions by means of which we achieve knowledge.[16] However, admitting the *natural* character of knowledge does not imply admitting its *naturalistic* character. Phenomenological apriorism provides us with an explanation of why we have to acknowledge the natural character of knowledge and the great importance of the natural sciences, and of why we, simultaneously, have to reject naturalistic interpretations of scientific theories. I take this explanation to mean the following: a) There is no ('purely intellectual') *a priori* knowledge rigorously dissociated from our experiential practice. b) There is no empirical knowledge which does not imply some *a priori* valid knowledge with regard to some of its moments and relations. In other words: a) All our knowledge and experience is part of nature. (To be sure, this is a mere tautology. Since we are humans inseparably bound to our natural existence, our knowledge and experience could, by definition, not occur independent of natural processes.) b) Nevertheless, this does not prove our knowledge and experience to be *exclusively* natural material. If it were, the meaning of phrases such as "I know that ..." would not be essentially different from the meaning of phrases like "It is drizzling."

What does "justification" and "self-justification" mean within the framework of transcendental phenomenology? What type of foundation does transcendental phenomenology support? In general, transcendental philosophies are occupied with analysing the basic concepts of our prescientific and scientific knowledge. According to Kant's and Husserl's idea of transcendental philosophy, adopting the reflective stance is compatible with the view that science proceeds *within its spheres of research* in an entirely autonomous style unimpeded by transcendental reasoning. The latter is interested in the most general conceptual structures which organize our experiences. It is directed to the most comprehensive understanding of what it means to have experiences of a certain range and order.

> Criticizing knowledge establishes a science which constantly and exclusively aims at elucidation of all kinds and forms of knowledge. Therefore, *it cannot use the results and existential assertions of any natural science;* these are to be kept in abeyance. All sciences are now considered as *science-phenomena.* Using them amounts to a faulty μεταβασις which results from a faulty though obvious *shifting of problems.* This shifting takes place between a psychological explanation of knowledge according to the methodical standards of the natural sciences on the one hand and analysing knowledge phenomena in terms of their intrinsic eidetic structures on the other hand. Hence, in order to avoid this shifting and to bear in mind the purpose of asking for eidetic possibilities, we require the *phenomenological reduction.*[17]

According to the phenomenological reduction we turn away from the ordinary objects of our acts of perceiving, thinking, remembering etc. and turn towards the intentional structure *noesis/noema* mediating our reference to these objects and states of affairs. Thereby we exclude any existential thesis with regard to the objects and states of affairs referred to as well as with regard to the acts of referring. Due to the phenomenological reduction, epistemology cannot be considered an integrated part of positive science. Any epistemological analysis requires a foregoing shift of the object domain, leaving behind scientific research fields. On the side of the subject this shift is reflected in a change of attitude. Although there is a kind of continuity between ordinary scientific investigation and a phenomenological critique of knowledge lying in *our attitude to change attitudes,* there is no continuity to be found with regard to the objects investigated. Analysing objects of diverse kinds (which is the task of empirical science) is not identical or 'continuous' with analysing the way objects appear to us (which is the task of philosophical science). This discontinuity allows the phenomenologist to reject the naturalistic circular-ity objection: A phenomenological critique of knowledge need not take any scientifically achieved knowledge as a premise of its own investigation. Neither does it make use of deductive-axiomatic explanations nor does it identify some privileged foun-dation in order to justify the objective validity of the knowledge we actually possess. Instead, it is interested in the meaning struc-ture lying beneath all our knowledge claims. The phenomeno-logical reduction does not establish a *tabula rasa* situation for analysing pure consciousness. Any such effort would amount to an arbitrary intervention altering the given that clearly would be incompatible with Husserl's methodological principles. Phenom-enologically understood, epistemology investigates the forms of (valid) intentional relations to objects. Such epistemology re-jects the following naturalistic anti-foundation thesis. Seeking philosophical foundations of scientific theories and prescientific experience can be defended without involving justificational foundationalism.

A phenomenological theory of intentionality modifies the idea of *adaequatio rei et intellectus*. According to Quine and other critics of external, 'extra-terrestrial' points of view, this idea is the stumbling-block of epistemology. But even if, as I take it for granted, phenomenology has to be relieved of the burden of a representationalistic idea of knowledge, i.e. knowledge as corre-spondence, there remains the problem of how to understand its self-justifying character. According to the naturalist's view the self-justification-thesis remains unfulfilled because of a *petitio principii* hidden in phenomenological epistemology. As sketched above, the allegedly presuppositionless investigation must pre-suppose its own feasibility, thereby anticipating right from the beginning what it pretends to show. Do we have to consent to this critique? Or is there anything wrong with it? Indeed, there is something wrong. First, transcendental phenomenology would be guilty of begging the question if and only if it pretended to justify knowledge (or 'objectivity') that is not within the scope of its descriptivist programme. (This descriptive analysis does not represent a theory of knowledge if the term 'theory' is taken in a rigid sense referring to a more or less complex system of deductively connected sentences.) Secondly, a phenomenological critique of knowledge does not aim at finding out *whether* valid knowledge is possible at all. Rather, it presents detailed descriptions of *how* it is possible to achieve knowledge with regard to different types of objects. Husserl's alignment of the epistemological task according to his descriptive program results from the nonsceptical starting-point of phenomenology. I shall return to the problem of scepticism later.

Making sense of Quine's epistemological naturalism requires reflecting on so-called 'external' and 'internal' approaches in epistemology. What do these terms mean when applied to the relation between transcendental philosophy and science? Here again, it turns out that naturalism harbours a double strategy naturalists often use. On the one hand, they interpret their own theoretical standpoint as insisting on the totally 'natural,' immanent character of all our epistemological reasonings. On the other, they classify traditional epistemology either as a kind of Cartesianism or transcendental externalism 'or as a combination of both (without recognizing the inconsistency of this combination).[18] However, transcendental phenomenology, although committed to a reflective stance, does not support any external or internal standpoint in terms of customary ways of employing these concepts. Instead, phenomenology establishes its own understanding of *internal* and *external* views. Within this theoretical framework, 'justifying knowledge' means to find the ground for an alleged assertion in some other knowledge. Only indirectly given *mittelbare* contents of knowledge can be justified in this sense. Conversely, this does not mean that directly (intuitively) given knowledge would be an unjustified positing. The difference concerns the specific way of how justification is achieved and the specific quality of being justified. In the case of intuitively given knowledge the justifying ground belongs to the phenomenon in question. It has its justifying ground within itself. Distinguishing purely signifying conscious contents and intuitive givenness does not introduce an internal/external distinction. According to a phenomenological view, both types of act qualities are internal if 'internal' means 'without having recourse to some transcendent entity existing (or being thought to exist) beyond the domain of possible experience.' Both types of act qualities are external if 'external' means 'being intersubjectively accessible without implying any privileged access or privacy hypothesis with regard to conscious contents.' When discussing naturalizing-programmes we have to be especially careful when using such concepts as *internal/external* or *subjective/objective*, whose meanings vary according to changing theoretical frameworks.

4. SELF-ABSOLUTISM

Since both naturalists and anti-naturalists feel free to handle autonomously their specific concepts and methods, it may be interesting to address what can be named the 'self-reflective aspect' of naturalism. The never-ending dispute regarding the promisory or futile character of traditional epistemology solidifies incompatible ideas of what epistemology is expected to do and of how it is related to scientific theories. Accordingly, we are also confronted with different criteria for judging the status and benefit of epistemological considerations. This meta-level debate cannot be rigidly separated from the object-level debate referring to particular problems of acquiring and sustaining knowledge. It is with reference to the meta-level debate that the superiority of a transcendental-phenomenological position becomes evident. Methodological naturalism, on the contrary, suffers from an inherent self-absolutism. A possible argument in favour of this thesis is as follows. Naturalism does not grant a special status to epistemology, leaving it all open to empirical investigations. Yet, no matter what empirical data are available in different fields of research and will be available in future, this nevertheless is of no help in considering epistemological (as well as metaphysical) questions. Empirical data do not tell us anything about how empirical and non-empirical moments build up our knowledge. Empirical data equally do not tell us how empirical and non-empirical dimensions of reality are related to each other. This being the case, a naturalistic epistemology that relies exclusively on empirically accessible information inevitably ends up absolutizing itself by rigidly excluding any possible 'non-natural' consideration. (Self-absolutism is nothing else but the other side of circularity.) However, naturalists can avoid this line of reasoning and they do. Quine explicitly warns against identifying naturalism with empiricism. When pressed in this matter, he probably would not hesitate to give up empiricism in favour of naturalism, although his relation to empiricism does not seem to be unambiguous.[19] Therefore, the self-absolutism argument, to be convincing, must be advanced independently of any attempt to associate naturalism and empiricism.

Self-absolutism results when a theory is incapable of restricting its sphere of validity. This incapability of self-limitation should not be considered in psychological terms, for instance as an unwillingness to adopt a critical stance with regard to one's own theoretical work. It is not a psychological matter. It is a matter of objective theoretical deficiency. Accomplishing the task of self-limitation requires transcending the object domain of the theory in question. Arguing from the point of view of a particular theory, one cannot justifiably assert that it must be possible to represent *everything there is* by means of the concepts and methods in question. Consequently, whenever a methodological naturalist wants to argue in favour of his theory, thereby rejecting idealistic or transcendental philosophies, he must have recourse to onto-logical naturalism, asserting that it lies exclusively within the competence of the natural sciences to state whatever could be part of our reality. A naturalist would tend to argue that whatever the future state of naturalistic methodologies will be, we grant them the exclusive right to decide what kind of objects are to be accepted as real objects. However, arguing this way leaves out the crucial point. No theory can demarcate its sphere of validity without, at least hypothetically, assuming a point of view that transcends the domain of objects in question. If we interpret our scientific work in terms of methodological naturalism, such a self-transcending movement is *per definitionem* excluded.[20] There is nothing beyond a 'natural' view. Self-absolutism is not a superficially-added interpretation of naturalized epistemology. It is the very heart of this programme.

According to traditional philosophical thinking, the idea of *reflection* refers to the ability of taking a critical view of one's own activities and their results. Reflecting upon one's theoretical or practical activities aims at objectifying and assessing them within a more comprehensive theoretical framework. In contradistinction to this traditional view a naturalistic epistemology, notwithstanding its efforts to reconstruct and interpret theoretically our scientific work, suffers from a particular lack of reflexivity. Let us consider an example to clarify this critique. Dirk Koppelberg, who has contributed to a better understanding of moderate naturalism, emphasizes the need to test and corroborate practically the idea of naturalization in the context of empirical research projects. In this connection he occasionally refers to recent investigations in the field of intentional psychology concerning the ascription of proposition attitudes, in particular the relation between first-person and third-person ascription. This reference to psychology is meant to illustrate the thesis that empirical research is relevant to the solution of genuinely philosophical questions which is one of the central theses of cooperative naturalism.[21] 'Cooperative naturalism' is a special type of methodological naturalism. It acknowledges that a philosophical epistemology uses concepts and norms and introduces principles and goals irreducible to those applied in scientific theories. Notwithstanding this half-way approach towards traditional epistemology, a cooperative naturalist denies both any genuinely philosophical method and that epistemology can claim any privileged status compared to ordinary scientific research. Although Koppelberg defends naturalism in a differentiated and subtle way, he too, begs the question. He tacitly presupposes there is no substantial difference between a philosophical interest we may take in the problem of ascription and the psychologically relevant questions. The latter especially concern the genetic priority of first-person or third-person ascriptions whereas the former, among other things, will be interested in discussing different criteria of distinguishing first- and third-person perspective and their epistemic appraisal. How could a psychological investigation ever demonstrate or make plausible that any nonpsychological interest we can have and actually do have concerning the ascription problem (for instance concerning questions of moral responsibility or the concept of free will) would be, in principle, illegitimate?[22] Or, as Koppelberg himself argues against Quine's programme: how could empirical science ever demonstrate or make plausible that there can be no other relevant data than empirical ones?[23]

The essence of methodological naturalism is the 'ism' that manifests itself in the movement to absolutism. According to this view, methodological naturalism is by no means a weak or conciliatory position apt to function as a mediator between naturalistic and anti-naturalistic theories. Nevertheless, scientific theories do not *necessarily* imply a naturalistic (or culturalistic or any -istic) commitment. Naturalism is a metatheoretical interpretation of scientific research programmes. The leading questions of this interpretation are eminently philosophical ones. They refer to the concepts of theory, experience, knowledge, justification, truth, objectivity, subjectivity, science, and reality. If self-absolutism is a general feature of dogmatic positions, then it goes without saying that ontological anti-naturalism cannot be considered a convincing alternative to naturalistic theories albeit in this case it is a question of an absolute positing of a higher, non-natural sphere of existence. Transcendental phenomenology will be able to avoid self-absolutism only if its self-justification implies some kind of self-limitation too. On the other hand, while eliminating any self-absolutism with regard to its own positing, it can explain why self-absolutism is unavoidably enclosed in naturalized epistemology.[24] Of course, naturalists would deny this and point out that the anti-naturalistic argument implies a *petitio principii*. Where does it lurk? The demand for excluding self-absolutism by means of a self-transcending operation already brings into play an anti-naturalistic idea of knowledge. Consequently, the naturalist may argue that, abandoning this idea, there is no motivation left for him to struggle to overcome self-absolutism. Do we then end up in a stalemate? Do we have to put anti-naturalism on a par with naturalism concerning self-absolutism? I do not think so. Whereas Quine dogmatically declares any attempt to discover *a priori* knowledge to be pointless, a transcendental phenomenologist does not diminish, restrict, or eliminate the research programmes of the natural sciences. Moreover, in introducing the above argument, the anti-naturalist does not tacitly impute her idea of knowledge to the naturalist's reasoning. Her argument, rather, is that a methodological naturalist cannot make sense of what he himself is doing without thereby either falling back on ontological naturalism (which is a much more vulnerable position than methodological naturalism) or hypothetically withdrawing the principles of *his own naturalizing project*. He cannot explain why *everything we experience*, in order to be qualified as part of our reality, had to be explained according to the concepts and methods of natural science. 'Natural' theories do not have a normative impact that could legitimate such an assertion. Thus, the naturalist is unable to explain why anyone should follow him in his naturalizing attitude.[25] If this is right, then methodological naturalism, on the one hand, is more dogmatic than it pretends to be and, on the other hand, suffers from a pragmatic self-contradiction.[26] It is a somewhat ironical turn that an argument against cultural relativism finally reflects on his own concept of naturalized epistemology. If there is no point of view outside natural science, then the naturalist must see his own point of view as absolute. He cannot proclaim naturalism without rising above it, and he cannot rise above it without giving it up.[27] I take it that the charge of *theoretical* self-refutation, when raised against naturalized epistemology, cannot prevail if the naturalist retires from the traditional epistemological project of digging for reasons and justifications and, therefore, could not even pretend to justify his own theoretical stance (albeit obviously underrates the sceptical twist of this strategy). Whether or not succeeds in avoiding both the theoretical self-refutation charge and an overtly sceptical overture of naturalism, his naturalized epistemology is, in any case, pragmatically self-contradicting.[28] There is no case for assuming that's naturalized epistemology and Husserl's transcendental phenomenology would represent incompatible but, nonetheless, equally good methodologies insofar as, after settling on one, we were unable to argue in its favour because any decisive factor we could bring into play would depend on one of the theories and, therefore, would anticipate its superiority. This is not the situation we are faced with. In the course of her investigation, a phenomenologist is led to reflect on possible limitations of her meaning constitution–analyses. Contrary to an epistemological naturalist she, moreover, is occupied with the presuppositions of her choosing a transcendental–phenomenological methodology. This is by no means an arbitrary reflection; rather, the transition from a natural attitude to a phenomenological attitude, and all its implications, is an explicit and important part of her analytical work, touching what Husserl calls a "phenomenology of phenomenology."

5. ANTI-NATURALISM, SCEPTICISM, AND REALISM

Contrary to his own conviction, a methodological naturalist cannot be content with accepting the circularity objection, which shows its destructive impact in connection with the dilemma of self-absolutism. When entering a phenomenological investigation, steering clear of the prejudices of our daily and scientific life, we realize that the sceptical intention to *doubt everything that can be doubted* rests on a thoroughly arbitrary decision.[29] Scepticism does not crop up as long as we are entangled in the different fields of our experience. Scepticism is not a natural attitude feature. *In this sense* a phenomenologist emphatically claims to analyse knowledge as a natural phenomenon. She claims to describe phenomena, not to 'reconstruct' them corresponding to changing theoretical goals. Starting with an arbitrarily-motivated sceptical doubt amounts to a complete misconception of phenomenological thinking. If we accept being bound to whatever is given (on varying conditions according to the essence of different types of experience and objects, respectively) we are not free to introduce unrestricted sceptical doubts. Universal scepticism cannot be refuted. Yet, we do not require such a refutation. Metaphysical Realism and scepticism are opposite poles going together well in terms of their mutual, 'dialectical' clinging together. Transcendental phenomenology breaks up this correlation. Consequently, it is not burdened with the "epistemological problem of the external world."[30]

Comparing methodological naturalism and methodological anti-naturalism with a view to their relation to scepticism reveals some particularly interesting aspects of the naturalism debate. This is due to the fact that our attitude towards scepticism essentially corresponds to our idea of epistemology. The naturalist is right in claiming that the ongoing process of scientific research should not be burdened with metaphysical scepticism. To be sure, if we use the term 'sceptical' in a non-specific and radically weakened sense, as merely indicating a critical handling of our hypotheses with a view to the incompleteness and corrigibility of all empirical knowledge, then scepticism is a 'natural' part of our scientific research. On the other hand, naturalists stress the fact that radical sceptical doubts will not occur unless we transcend our scientific work-in-progress in order to look for some *a priori* guarantee of our knowledge. If we free ourselves from the 'invisible' threat of universal scepticism which has been theoretically overestimated for centuries, so the naturalist may argue, we realize that we actually do not need a Cartesian epistemology helping us to get rid of scepticism. Scepticism in its proper sense arises whenever we naïvely accept the so-called *thing-in-itself* which is said to be the real object of our knowledge. But, as Quine rightly throws in, the very positing of a thing-in-itself is in need of justification. This justification is still lacking.

Husserl undoubtedly would agree with Quine that we are not obliged to doubt everything which could possibly be doubted. Such an all-comprising hyperbolic doubt would only be legitimate if we took for granted a thing-in-itself as the proper object of knowledge. However, we do not need to refer to this problematic idea when trying to understand how it is possible to know something about the world we experience. Of course, Quine's naturalism's and Husserl's critique of a thing-in-itself is based on radically different conceptions of the immanent character of our knowledge. Whereas Quine directly attacks the conceptual distinctions between transcendence and immanence, transcendent and transcendental, and between immanent and transcendental,[31] Husserl, on the contrary, conceives of his immanent descriptive analyses as a transcendental investigation excluding any reasoning which transcends the given phenomena. Although the naturalist is right in maintaining that scepticism does not play a crucial and fatal role within the scope of our daily scientific work, this certainly does not imply that scepticism disappears altogether[32] or could be paralysed by considering it as an offshoot of science.[33] The naturalist dumping strategy does not allow us to face the sceptical challenge in a satisfactory way. It does not prevent universal scepticism from reappearing. Pragmatically, it advises us to ignore any thoroughgoing sceptical doubt. Without this foregoing taming of scepticism, Quine's proposal to reconcile scepticism with his vague idea of "robust realism" would appear totally implausible. Coming to terms with scepticism in a more profound way *without having recourse to a Cartesian programme of refuting scepticism* demands that we investigate what is enclosed in the meaning content of our actual knowledge claims. Following the phenomenological descriptions of different types of experience, and accordingly, different types of objects, we realize that scepticism will be successful until we radically challenge the basic ideas of our epistemological reasoning. We can unsettle the principle of universal scepticism (doubting everything which can be doubted) by unsettling the idea that our knowledge needed some *fundamentum inconcussum* and resulted from some mysterious corresponding relation because otherwise our knowledge would not be real knowledge and our 'immanent' ideas would fail to grasp 'real' things.

Notes

1 See Hilary Kornblith, ed., *Naturalizing Epistemology*, 2d ed. (Cambridge MA, London: The MIT Press, 1994); Peter A. French, Theodore E. Uehling, and Howard K. Wettstein, eds., *Philosophical Naturalism*, Midwest Studies in Philosophy, vol. 19 (Notre Dame, IN: University of Notre Dame Press, 1994); Geert Keil and Herbert Schnädelbach, eds., *Naturalismus: Philosophische Beiträge* (Frankfurt: Suhrkamp Verlag, 2000); Susan Haak, "Naturalism Disambiguated," *Evidence and Inquiry: Towards Reconstruction in Epistemology* (Oxford, Cambridge: Blackwell Publishers, 1993), 118–138; Winfried Löffler, "Naturalisierungsprogramme und ihre methodologischen Grenzen," in *Der neue Naturalismus – eine Herausforderung an das christliche Menschenbild*, ed. Josef Quitterer and Edmund Runggaldier, 30–76 (Stuttgart, Berlin, Köln: W. Kohlhammer Verlag, 1999); Thomas M. Seebohm and J. Gutenberg, "Psychologism Revisited," in *Phenomenology and the Formal Sciences*, ed. Thomas M. Seebohm, Dagfinn Føllesdal, and Jitendra Nath Mohanty, 149–182 (Dordrecht, Boston, London: Kluwer Academic Publishers, 1991).

2 From a phenomenological point of view, I approve of the idea that the naturalizing-epistemology debate is decided in the field of intentionality: "Ob die Erkenntnistheorie sich mit naturalistisch respektablen Phänomenen beschäftigt, entscheidet sich nicht in der Erkenntnistheorie, sondern am Erfolg oder Mißerfolg einer Naturalisierung des Intentionalen. Kurz: Erkenntnistheorie ist bei Strafe des Themenwechsels intentionalitäts-präsupponierend. Untersuchungen, die über Naturalisierbarkeit entscheiden sollen, müssen aber intentionalitätsproblematisierend sein." Geert Keil, "Naturalismus und Intentionalität," in *Naturalismus: Philosophische Beiträge*, ed. Keil and Schnädelbach, 201. Nevertheless, this view *presupposes* a certain idea of knowledge and of what epistemology has to achieve.

3 See W.V.O. Quine, "Naturalismus-oder: Nicht über seine Verhältnisse leben," in *Naturalismus: Philosophische Beiträge*, ed. Keil and Schnädelbach, 113–121.

4   A paradigmatical misinterpretation of phenomenology is found in Richard
    Rorty, "Epistemological Behaviorism and the De-Transcendentalization of
    Analytic Philosophy," *Neue Hefte für Philosophie* 14 (1978): 114–142. For
    Rorty, the outcome of the phenomenological reduction is some sort of
    privileged representation that presupposes a picture idea of knowledge
    ('accurate representing') and, therefore, a concept of truth as correspon-
    dence. Phenomenology, whose distinctive feature would be the quest for
    certainty, is said to defend and promote the 'Mirror of Nature' model,
    thereby continuing the old 'Platonic project.' Rorty flatly contradicts Husserl
    by claiming that phenomenology attempts to establish philosophy as a
    rigorous science *without taking into account the relation between subject and
    object*. When speaking of a "de-transcendentalizing" movement, Rorty
    describes a certain phase in the development of analytic philosophy. Nev-
    ertheless, he espouses a more general view of the aims of transcendental
    philosophy. But this distorted picture of phenomenology must be cor-
    rected. See S. Rinofner-Kreidl, *Edmund Husserl: Zeitlichkeit und Intentionalität*
    (Freiburg and München: Karl Alber Verlag, 2000).

5   See Husserl's analogous arguments against logical psychologism pointing
    to a metabasis in the field of logic. Edmund Husserl, *Logische Unter-
    suchungen. Erster Theil: Prolegomena zur reinen Logik, Husserliana* (den Haag:
    Martinus Nijhoff, 1975), 18: chapter two.

6   See Edmund Husserl, *Philosophie als strenge Wissenschaft* (Frankfurt: Vittorio
    Klostermann, 1965), 23–64; Edmund Husserl, *Ideen zu einer reinen
    Phänomenologie und phänomenologischen Philosophie, Drittes Buch: Die
    Phänomenologie und die Fundamente der Wissenschaften, Husserliana* (den
    Haag: Martinus Nijhoff, 1952), 5: 72–75; Edmund Husserl, *Die Idee der
    Phänomenologie. Fünf Vorlesungen, Husserliana* (den Haag: Martinus Nijhoff,
    1950), 2: 6 ff., 20–26.

7   The following summary concentrates on the basic character of Quine's
    naturalized epistemology. I leave aside more special issues (e.g. holism,
    ontological indeterminateness, behaviourism) which should be discussed
    separately.

8   Note that some version of the circularity objection is advanced by thinkers
    who are by no means antagonistic to anti-naturalistic philosophical criti-
    cism. See for instance Nelson's famous argument that any search for crite-
    ria intended to guarantee the objective validity of our knowledge necessarily
    gets entangled in an infinite regress, a circular structure or a contradiction.
    Leonhard Nelson, "Über das sogenannte Erkenntnisproblem," (1908) and
    "Die Unmöglichkeit der Erkenntnistheorie," (1911), Leonhard Nelson,

Geschichte und Kritik der Erkenntnistheorie, Collected Writings (Hamburg: Felix Meiner, 1973), 2: 59–393, 459–483.

9 "We are after an understanding of science as an institution or process in the world, and we do not intend that understanding to be any better than the science which is its object. This attitude is indeed one that Neurath was already urging in Vienna Circle days, with his parable of the mariner who has to rebuild his boat while staying afloat in it." Quine, "Epistemology Naturalized," *Ontological Relativity and Other Essays* (New York and London: Columbia University Press, 1969), 84.

10 Quine, "Epistemology Naturalized," 75ff.

11 See Quine, "Two Dogmas of Empiricism,"in *From a Logical Point of View: 9 Logico-Philosophical Essays*, W.V.O. Quine, 2d ed. (New York and Evanston: Harper & Row, 1963), 20–46. Whether the above quotation and similar passages in Quine's work point to either discarding or reformulating the idea of normative epistemology is controversial. I side with the former view. There is nothing interestingly normative left after Quine's "technologizing" of epistemic norms. For this discussion see Richard Foley, "Quine and Naturalized Epistemology," *Midwest Studies in Philosophy* 19 (1994): 243–260.

12 "There is no *tabula rasa*. We are like mariners who have to rebuild their ship while sailing on the sea without ever being able to take it apart in a harbour and reconstruct it using solid components. Only metaphysics can disappear without remainder" (my translation). "Es gibt keine tabula rasa. Wie Schiffer sind wir, die ihr Schiff auf offener See umbauen müssen, ohne es jemals in einem Dock zerlegen und aus festen Bestandteilen neu errichten zu können. Nur die Metaphysik kann restlos verschwinden." Otto Neurath, "Protokollsätze," *Gesammelte philosophische und methodologische Schriften, Band* 2 (Wien: Hölder-Pichler-Tempsky, Verlag, 1981), 579. I leave out Neurath's reference to the protocol-sentence idea. The nonexistence of a *tabula rasa* means that we cannot indicate "pure" protocol-sentences. No part of our system of scientific judgments can be made entirely precise and certain.

13 The naturalistic strategy of minimizing reasonable alternatives can be discovered, for instance, in opposing sensual perception and extra-sensual perception (as introduced in parapsychological writings). By rejecting the latter the naturalist suggests that there could not be any other type of experience than sensual experience. See Gerhard Vollmer, "Was ist Naturalismus?" in *Naturalismus: Philosophische Beiträge*, ed. Keil and Schnädelbach, 63ff.

14 Moreover, eidetic phenomenological judgments show an ambivalent character if considered in terms of the analytic-synthetic distinction. See Rinofner-Kreidl, *Edmund Husserl: Zeitlichkeit und Intentionalität*, 98–107, 139–150.

15 Dirk Koppelberg, "Why and How to Naturalize Epistemology," in *Perspectives on Quine*, ed. Robert B. Barrett and Roger F. Gibson (Oxford: Basil Blackwell, 1990), 210.

16 It is puzzling how enthusiastic naturalists are about trivial facts. For instance, consider Vollmer's remark "Verstehen gelingt nur mit Hilfe unseres Gehirns, also eines natürlichen Organs." Vollmer, *"Was ist Naturalismus?"* 64. Who would deny this?

17 "Ist also ... die Erkenntniskritik eine Wissenschaft, die immerfort nur und für alle Erkenntnisarten und Erkenntnisformen aufklären will, so kann sie von keiner natürlichen Wissenschaft Gebrauch machen; an ihre Ergebnisse, ihre Seinsfeststellungen hat sie nicht anzuknüpfen, diese bleiben für sie in Frage. Alle Wissenschaften sind für sie nur Wissenschaftsphänomene. Jede solche Anknüpfung bedeutet eine fehlerhafte μεταβασις. Sie kommt auch nur zustande durch eine fehlerhafte aber freilich oft naheliegende Problemverschiebung: zwischen psychologisch naturwissenschaftlicher Erklärung der Erkenntnis als Naturtatsache und Aufklärung der Erkenntnis nach Wesensmöglichkeiten ihrer Leistung. Es bedarf also, um diese Verschiebung zu meiden und beständig des Sinnes der Frage nach dieser Möglichkeit eingedenk zu bleiben, der *phänomenologischen Reduktion"*(my translation). Edmund Husserl, *Die Idee der Phänomenologie: Fünf Vorlesungen* (den Haag: Martinus Nijhoff, 1950), 6.

18 See for instance P.T. Sagal, "Epistemology De-Naturalized," *Kant-Studien* 69 (1978): 98ff.

19 See Dirk Koppelberg, "Was ist Naturalismus in der gegenwärtigen Philosophie?" in *Naturalismus: Philosophische Beiträge*, ed. Keil and Schnädelbach, 69–72.

20 This type of argument equally applies to (methodological) culturalism *if defending this position means to assert that there is nothing beyond a cultural view.* The argument is directed at any dogmatic interpretation of theoretical frameworks. Thus, I agree with Putnam's remark that "cultural relativism ... is one of the most influential – perhaps the most influential – forms of naturalized epistemology extant, although not usually recognized as such." Hilary Putnam, "Why Reason Can't Be Naturalized," in his *Realism and Reason: Philosophical Papers* (Cambridge: Cambridge University Press, 1983), 3: 235.

21 See Dirk Koppelberg, "Was ist Naturalismus in der gegenwärtigen Philosophie?" 91.

22 There is a parallel case in methodology. How can the distinction between empirical and transcendental questions be introduced within the range of empirical theories? How can empirical science ever demonstrate this distinction to be, in principle, invalid? It certainly is consistent to refute the empirical/transcendental distinction *if one takes the absolutizing-science strategy*. But can one consistently follow this absolutizing strategy (see below)? Accepting the empirical/transcendental distinction, vice versa, implies the rejection of self-absolutist interpretation of scientific theories. See Edmund Husserl, *Die Idee der Phänomenologie: Fünf Vorlesungen*, 22ff.

23 See Bonjour's treatment of the question of whether a naturalist can give any argument for excluding *a priori* justification. I categorically agree with the author on this matter. Laurence Bonjour, "Against Naturalized Epistemology," *Midwest Studies in Philosophy* 19 (1994): 291–295.

24 These assertions require more detailed explanation than can be given here. See my work cited in n.4 and n.29. In this context we face a deep and controversial problem, namely, the relation between immanence and transcendence and its conceptualization within different philosophical paradigms. Some naturalists turn the tables accusing anti-naturalists of being unable or unwilling to give any plausible explanation of why we are limited in our knowing abilities. See Richard Rorty, "Epistemological Behaviorism and the De-Transcendentalization of Analytic Philosophy," *Neue Hefte für Philosophie* 14 (1978): 131ff. "But what exactly do we want? Advice for any reasoning being – for 'reason itself' – that would be good no matter what the world is like? Or advice for limited creatures like ourselves that would be effective in the actual world?" Philip Kitcher, "The Naturalists Return," *The Philosophical Review* 101 (1) (1992): 64.

25 This meta-level normativity problem should not be confused with object-level discussions of the descriptive-normative distinction. See Quine, "Naturalismus – oder: Nicht über seine Verhältnisse leben," 122.

26 This charge is rejected in Dirk Hartmann and Rainer Lange, "Ist der erkenntnistheoretische Naturalismus gescheitert?" in *Naturalismus: Philosophische Beiträge*, ed. Keil and Schnädelbach, 154–156. Here, the objection is raised referring to the naturalist's inability to give a naturalistic explanation of those acts by means of which he expresses his own theory. Contrary to this, the above argument refers to the impossibility of naturalistically *arguing* in favour of methodological naturalism.

27 "Truth, says the cultural relativist, is culture-bound. But if it were, then he,

within his own culture, ought to see his own culture-bound truth as absolute. He cannot proclaim cultural relativism without rising above it, and he cannot rise above it without giving it up." Quine, "On Empirically Equivalent Systems of the World," *Erkenntnis* 9 (1975): 327ff.

28 For this objection see Bonjour, 1994, especially 296ff. Formerly, the thesis that a naturalistically interpreted natural science would be theoretically self-contradicting is, for instance, advocated in Leonhard Nelson. "*Ist metaphysikfreie Naturwissenschaft möglich?*" Leonhard Nelson, *Die kritische Methode in ihrer Bedeutung für die Wissenschaft* (Hamburg: Felix Meiner Verlag, 1974), 233–281. Concerning Nelson's "Widerspruch der empiristischen Grundvoraussetzung," 268–271, compare Husserl's largely akin reasoning in the first volume of his *Logical Investigations*. Empiricism is, in the end, identical with scepticism. Any (theoretically relevant) scepticism is self-contradictory. See Edmund Husserl, *Logische Untersuchungen, Erster Theil: Prolegomena zur reinen Logik* (den Haag: Martinus Nijhoff, 1975), chapters 32–38.

29 I tackle this problem in "Die Entdeckung des Erscheinens. Was phänomenologische und skeptische Epoché unterscheidet," *Allgemeine Zeitschrift für Philosophie* (forthcoming).

30 Regarding transcendental phenomenology, I do not agree with the view that "idealism does not radically replace naturalism. In order to overcome naturalism radically, one also has to reject the principle of immanence and the epistemological problem of the external world which is implied by it." Herman Philipse, "Transcendental Idealism," in *The Cambridge Companion to Husserl*, ed. Barry Smith and David Woodruff Smith (Cambridge: Cambridge University Press, 1995), 300. According to Philipse, and contrary to the interpretation espoused above, Husserl's phenomenology is part of the *traditional* foundationalistic programme (in its classical and modern types).

31 See "turning transcendental questions into immanent ones. In the end, the contrast between transcendental and immanent disappears completely just as the exterior-interior contrast disappears in a Moebius strip in which only one surface remains, as can be seen when we try to paint it all over." Paul Gochet, Quine's Epistemology," in *Pragmatische Tendenzen in der Wissenschaftstheorie*, ed. Herbert Stachowiak (Hamburg: Felix Meiner, 1995), 135.

32 Of course, one will be inclined to believe in that *if* one agrees that naturalistic meta-epistemology should (and could) take the place of normative epistemology and that all traditional answers to scepticism fall under normative epistemology. Yet, the whole quarrel is about whether or not

these views stand scrutiny. See Richard Fumerton, "Skepticism and Naturalistic Epistemology," *Midwest Studies in Philosophy* 19 (1994): 324, 333, 338ff. A great part of the Anglo-Saxon naturalizing epistemology debate is concerned with the problem of normative epistemology. It deals with the consequences we should or should not draw from the fact that the way we *ought* to arrive at our beliefs (often) significantly differs from the way we actually *do* arrive at our beliefs. In any case, calling into question sceptical strategies from a phenomenological point of view is a descriptive programme. Phenomenology does not pretend to tell us which of our beliefs about the world are justified beliefs.

33 See W.V.O. Quine, "The Nature of Natural Knowledge," in *Mind and Language*, ed. Samuel Guttenplan (Clarendon Press: Oxford, 1975), 67ff.

CHAPTER THREE

Denis Fisette

Erläuterungen: Logical Analysis vs. Phenomenological Descriptions

———

> *Die Bedeutungen von Urzeichen können durch Erläuterungen erklärt werden. Erläuterungen sind Sätze, welche die Urzeichen enthalten. Sie können also nur verstanden werden, wenn die Bedeutungen dieser Zeichen bereits bekannt sind.*
>
> L. Wittgenstein

Husserl, a trained mathematician, just like Frege and Bolzano, and student of two of the most notable scholars of that field, Kronecker and Weierstrass, had first-hand knowledge of his contemporaries' scientific work. Although his contribution to mathematics as such remains modest, one would be wrong to minimize the importance of formal and natural sciences within Husserl's philosophical itinerary. For instance, his project of a universal *mathesis* and the articulation of his doctrine of definite manifolds were Husserl's response to mathematical problems, namely, those of imaginary numbers, and are among the few ideas to which Husserl remained faithful until the end of his career. Yet, judging upon the work that he published during his life, his interest in the sciences, in particular natural science, is mainly philosophical. Husserl was not so much concerned with contributing to the actual progress of science as with spelling out his own relationship to the mainstream philosophical position of the time known

———

as naturalism. Inasmuch as Husserl is not interested in natural science *per se* but in the philosophical claims of those who practice it, it is precisely the latter concern which may be said to be at work when he criticizes the naturalistic beliefs in which the rising psychology, experimental psychology, and psychophysics in particular, were embedded. This form of naturalism is particularly radical. The attitude it adopts towards other philosophical positions is comparable to that of chemistry towards alchemy, that is, it relegates philosophical concepts drawn from common sense to the rank of chimera and fiction, and seeks to replace them with concepts authorized by science.[1] Naturalism is Husserl's target in many of his works, particularly in *Krisis* where it is held responsible for what he calls 'objectivism': a prejudice supposedly inherited from Galileo, which maintains that "the exact sciences of nature guarantee absolute metaphysical knowledge." If it is indeed the case that the natural science of psychology claims to form the scientific foundation for logic, the sciences of the mind, and even metaphysics itself, then naturalism is also Husserl's target in *Philosophy as a Rigorous Science*.[2] Furthermore, if we admit that the epistemological and metaphysical presuppositions of logical psychologism epitomize the very principles of philosophical naturalism, Husserl's critique of naturalism is perhaps not wholly unconnected to his critique of logical psychologism in *Prolegomena*. Psychologism would thus be an epistemological thesis that ascribes a foundational role to scientific psychology.

Husserl and Frege's arguments against logical psychologism are known, but what about alternatives? Any critique of this form of psychological naturalism is confronted by the following. One must either seek to dissociate logic from any epistemological considerations, or attempt to work out a theory of knowledge that will be sufficiently radical to elude anti-psychologistic objections and resume its foundational role. According to Dummett, whilst Frege, inasmuch as he substitutes a logical analysis of language for epistemology, belongs to the first group; Husserl, who is seeking to revive the theory of knowledge through phenomenology, belongs to the second. This divergence emerges clearly from their opposition concerning the analysis of the concept of number and of primitive logical concepts. I will identify what is philosophically at stake in the disagreement between Husserl and Frege, and will assess the validity of their respective criticism of psychological naturalism.

1.

Let us start from what has been called the Fregean reading of phenomenology, which has two aims: 1) to show that Frege, Husserl's contemporary, could do for phenomenology what he had done for twentieth century analytical philosophy and 2) to turn Frege's work into the chief source of inspiration for Husserl's phenomenology. This approach to phenomenology has been dominant since the end of the 1960s and has proved to be extremely fruitful. It has made it possible to approach phenomenology from a new angle and to assess it in the light of contemporary problems. It has not exhausted all its weapons; on the contrary, just as any other venture of this type, this approach will remain effective only if its proponents recognize its limitations – in particular, that its scope does not stretch beyond certain aspects of the theory. Husserl himself clearly drew the limits in *Philosophy of Arithmetic,* in which he criticizes Frege's account of the origin of the concept of number. Despite many changes in Husserl's thought, the rationale of his critique of Frege's analyses and definitions is also what led him, some ten years later, to entrust this investigation to phenomenology.

The Fregean reading of phenomenology emphasizes three aspects of Husserl's relation to Frege. The first concerns the influence Frege's review of *Philosophy of Arithmetic* might have had on Husserl's pre-phenomenological antipsychologistic turn. There are no documents, no explicit testimony of Husserl, to corroborate this thesis, but we have grounds enough to believe that Husserl could not remain impassive to Frege's otherwise constructive critique in his review of *Philosophy of Arithmetic*.[3] The second aspect is that one of Frege's reproaches rests on the notorious *Sinn/Bedeutung* distinction, a distinction which will end up playing a significant role in Husserl's phenomenology in *Logical Investigations*. The fact that this distinction is also to be found in Bolzano, Lotze, Twardowski, and even Kerry (as the most ardent proponents of the first two theses have readily acknowledged), does not constitute an objection to the idea that Frege's astute analyses might have been very useful in the clarification of the corresponding distinction in Husserl.[4] But this sole distinction, as the arguments of *Prolegomena* show, does not justify Husserl's antipsychologistic turn. The third aspect is that there is a family resemblance between Husserl's arguments in *Logical Investigations* and those we find scattered throughout Frege's work, particularly in his review of *Philosophy of Arithmetic* and the preface of the *Grundgesetze*. The argument, in a few words, concerns the ideal nature of laws, principles, and propositions of logic and of their incompatibility with the laws, principles, etc., of psychology conceived as an empirical science. Despite the interpretation of logical psychologism which has prevailed in the post-Fregean tradition, and in particular in Carnap, and according to which psychologism is the attempt to reduce a normative science to a descriptive one, I feel that the third thesis is hardly questionable, but I do not intend, at least for now, to enter this debate.[5]

In conjunction with this issue is the nature of the division between the analytic and continental philosophical traditions, which arose after Frege and Husserl and dominated twentieth century philosophy. An important thesis, put forth by Dummett, states that the respective philosophies of the two philosophers were at the time so alike that nothing in the philosophy of *Logical Investigations* hinted at such a division.[6] The semantic theme Husserl favoured in *Logical Investigations* does indeed bring him remarkably near Frege and the tradition he inspired, from

Wittgenstein to Quine. According to Dummett, the division settled in soon after the publication of *Logical Investigations* and seemed to ensue from the philosophical status Husserl conferred on his phenomenology, a status that would bring him seriously close to Kantian idealism.

Two observations concerning Dummett's historical approach are indispensable. First, his narrative does not go beyond this historical segment which was indeed dominated by post-Fregean philosophy, that is, by logical positivism and ordinary language philosophy. But what is there to say about the strong return of naturalism and of psychologism in the philosophy of the last thirty years, which has been dominated by the philosophy of mind and cognitive sciences? It makes sense to say that since Quine, the field of philosophical psychology has been restored and that the expedient of this restoration was a critique of the philosophy inherited from Frege. Moreover, this partiality of contemporary philosophy for consciousness and mental phenomena brings it nearer to phenomenology as practiced by the Brentano School and by the early phenomenologists than to Frege's philosophy. I will return to this.

Second, what about phenomenology itself, and above all, what about the phenomenology of *Logical Investigations*? Of course, one is justified in emphasizing the likeness between the two philosophers on issues as important as the theory of meaning and the arguments against psychologism, but the philosophical significance of the *Logical Investigations* goes far beyond that, and a one-sided Fregean reading of phenomenology is likely to obliterate other ideas which are present in this work and which we hold to be responsible for the orientation phenomenology will take soon after its publication. This interpretation of phenomenology is mistaken; however, a reflection on phenomenology that accounts for current philosophical debates, will inevitably encounter this question. Are phenomenologists better equipped than Frege and his followers to explain the nature of consciousness and phenomenal experience?[7]

2.

The first obstacle on which our investigation stumbles, which is also the reason why the above described perspective has largely been brushed aside by commentators, concerns the very sense of phenomenology, be it in *Logical Investigations* where it is conceived as descriptive psychology or in *Ideen* I, where it nurses transcendental philosophy. For now, I will focus on *Logical Investigations*. In other words, the first problem concerns the tension which exists in this work between *Prolegomena's* arguments against psychologism and the psychological theme which occupies as important a place in the subsequent investigations, namely in the Fifth *Logical Investigation* where the central topic is the intentionality of acts. Moreover, phenomenology, which investigates the origins of fundamental logical concepts and laws,[8] depicts itself as descriptive psychology.[9] An ill-disposed Fregean might read this as evidence that Husserl did not benefit from Frege's review, which condemns *Philosophy of Arithmetic's* confusion between *'Gedanken'* and *'Vorstellungen,'* that is, between the psychological act and its logical content or meaning. This confusion would also be at the source of Husserl's psychologizing the concept of numbers in his first book. According to Frege, this would obliterate the distinction between the objective and ideal features of numbers and their subjective representation, the latter being in nature private and falling under the "psychological laws of association."[10]

But one of the principles which has continually guided Frege's philosophy, namely the first principle in *Foundations of Arithmetic*, is that one must clearly distinguish what is psychological from what is logical, the subjective from the objective. There is much to say about the sense ascribed to psychology and subjective representations in this principle, but we may assume that it refers to introspective psychology. At any rate, Frege's apprehension towards psychology raises two important questions. First, do the studies of the second volume of *Logical Investigations* on psychological themes represent a relapse into the form of psychologism condemned in *Prolegomena*? Second, does Husserl's conception of phenomenology as descriptive psychology differ from the type of psychology Frege is seeking to expel from the field of logic? I will briefly respond to the first question, since I treat it in detail elsewhere, before I concentrate on the second.

The phenomenological theme of *Logical Investigations* reveals an important difference with Frege. This difference concerns the philosophical relevance of the psychological field of investigation. As the principle of *Grundlagen* above mentioned clearly indicates, not only does Frege defend an antipsychologistic position with respect to all questions relative to the foundations of logic and mathematics, but he also believes that philosophy should expect nothing from psychology. The confusion between these two theses in the post-Fregean tradition is the source of a gap between philosophy and psychology. In contrast, while remaining faithful to a form of antipsychologism or another, phenomenology has always maintained ties with psychology, up until *Krisis* where (transcendental) phenomenology is clearly identified with phenomenological psychology, a novel version of descriptive psychology. Under these conditions, what can we say about the relation between logic and psychology in *Logical Investigations*? It seems to me that our answer must take into account the bilateral struggle Husserl is leading with his critique of logical psychologism and the double-sided motive which underlies *Logical Investigations*: on the one hand, the psychological motive which concerns the subjective dimension of the act of thinking and which is the object of the last two *Logical Investigations*; and, on the other hand, the logical motive, which concerns the objective and ideal nature of meaning and the reference to objects.[11] The logical motive criticizes the conception of logic as a praticonormative discipline and leads to the idea of a pure logic as sketched in the last chapter of *Prolegomena*. The psychological motive criticizes the foundational claims of physiological and experimental psychology; Husserl ascribes this foundational role to his own phenomenology. These two motives constitute the framework of *Logical Investigations* and account for the fact that Husserl's concern is not merely limited to sustaining the "objectivity of the content and object of knowledge" as it is the case in Frege, but also and above all to accounting for its relation with "the subjectivity of the act of knowledge."[12] This is the first important difference between Husserl and Frege.

But here arises the second question, namely whether Husserl's conception of phenomenology as descriptive psychology eludes Frege's critique and differs from the psychology the latter is seeking to drive out of the field of logic and philosophy in general. As it is the case in Brentano from whom Husserl borrows the expression "descriptive psychology" and which refers, in *Psychology from an Empirical Standpoint*, to the "science of mental phenomena," it departs from explicative or inductive psychology inasmuch as the latter espouses the method of natural sciences. This divergence is essential to understanding the sense Husserl ascribes to his phenomenology as well as the actual target of his antipsychologism. As one of the passages of his review of Palágyi's book, *The Conflict of the Psychologists and Formalists in Modern Logic*,[13] shows, the struggle against psychologism in *Prolegomena* is hardly a "struggle against the psychological foundation of logical methodology or against the elucidation, by descriptive psychology, of the origin of logical concepts, but merely a struggle against an epistemological point of view." It is clear that psychology is concerned only insofar as it serves a radical form of philosophical naturalism, a position which seems to be ascribable to Wilhelm Wundt as well as to John Stuart Mill. But what about descriptive psychology and its relation to phenomenology? Our hypothesis is the following: Husserl's interest in descriptive psychology applies less to psychology as such than to its status as a descriptive science. This is confirmed by Husserl in many texts, e.g. in his 1903 reviews of Palágyi's and Elsenhans's books, in the draft to a preface to the second edition of *Logical Investigations*, and in particular, in the second edition of the introduction to *Logical Investigations* where he reminds us that phenomenology is not descriptive psychology in the 'old' sense of the term.[14] By 'old' we understand the 1874 Brentanian version, which refers to a method grounded in introspection which Husserl criticizes in length in an appendix to the Sixth Logical Investigation. This critique is precisely about the sense Brentano ascribes to the notion of description.[15]

In a few words, Husserl challenges Brentano's criterion for the distinction between mental and physical phenomena, i.e. the notion of evidence; the problem is that Brentano did not make the distinction, otherwise inevitable for a phenomenologist, between adequate and apodictic evidence. So the phenomenological purpose of evidence is not to be understood in terms of (clear and distinct) perception, but in terms of description. Under a description and, more precisely, under a descriptive criterion for the distinction between two types of phenomena, one must understand a negative criterion which implies "no supposition whatsoever from an epistemological standpoint," no "presuppositions with respects to metaphysical reality." The adjective "descriptive" is thus opposed to what is "oriented by the supposed data of the transcendent world" and the purely descriptive characterization of a phenomenon means oriented towards the true "*Gegebenheiten*" of phenomena.[16] Hence, Husserl's main objection to Brentano is that the latter grounded his psychology on empirical descriptions which concern "the real states of animated beings and natural reality." It is otherwise with phenomenological descriptions:

> Phenomenology, however, does not discuss states of animal organisms (not even as belonging to a possible nature as such), but perceptions, judgments, feelings *as such*, and what pertains to them *a priori* with unlimited generality, as *pure* instances of *pure* species, of what may be seen through a purely intuitive apprehension of essence, whether generic or specific. [17]

In the first instance, phenomenological descriptions do not concern the physical properties of worldly objects but the manner in which we experience them, what Husserl here calls experience in the strict sense, the experience as such or its meaning. In other words, phenomenology analyses the modes of givenness of things and not the things that are given. This is analogous to Frege's analysis of *Sinn* in terms of *Gegebenheitsweise*, i.e. modes of givenness.

It is thus obvious that the double motive mentioned above corresponds to the notorious distinction presented in the last chapter of *Prolegomena* between ontological and nomological sciences. Pure logic, which Husserl conceives as Leibniz did – as a *mathesis universalis* – is just like geometry: a nomological science whose unity is ensured by its fundamental laws. The latter Husserl interprets, shortly after *Logical Investigations*, in Hilbert's sense, in terms of axiomatic system.[18] On the other hand, regional sciences such as anatomy, natural history, or psychology, which are material sciences, owe their unity and their content not to laws but to their object or respective domain of objects. It is precisely the latter that Husserl characterizes as descriptive "since the unity of description is fixed by the empirical unity of the object or the class, and it is this descriptive unity which, in the sciences here involved, determines the science's unity."[19] Now, as we have just seen, phenomenology is itself a descriptive science. Among the descriptive sciences we must distinguish between those (like Brentano's descriptive psychology) concerned with the objects belonging to the class of real events such as mental phenomena, and those (like phenomenology) which are philosophically neutral, that is, that bracket the presuppositions of the former regarding metaphysical reality. Although the field of study of phenomenology overlaps that of descriptive psychology, it is only interested in the modes of givenness. These may in return be understood in terms of essences, as is the case in *Logical Investigations*, or in terms of meaning or noematic sense, as in Husserl's later works. It is thus in this sense that phenomenology is not a descriptive psychology in the old sense of the term.

3.

But why should we give so much importance to psychologism in a work that understands itself as prolegomena to pure logic? In other words, why should we not be content with the logical motive and the highlighting of a pure logic which, as we know, is a nomological science and, as such, autonomous or independent of any regional science such as psychology? Should we see it as self-criticism as Husserl suggests in the preface to *Logical Investigation*'s first edition, in the last chapter of *Prolegomena*, and in many other texts? Is it a matter of better delineating the empiricist conception of logic, which prevailed at the time, and using it against the Kantians in the disputes that oppose Husserl to them? In fact, Husserl grants to psychologists that practico-normative logic requires technical prescription which is especially "adapted to human nature."[20] For instance, when dealing with methodological tools (abacus, telescope, etc.) it is necessary to take "mental processes" into account. But this is no exception since all logical concepts such as truth, judgment, inference, etc. have a "psychological origin" and thus refer to mental experiences.[21] However, this psychological aspect of all concepts pertaining to the technology of logic does not exhaust their theoretical content. This is the implication of Husserl's critique of psychologism. Indeed, every logical concept, as for instance the concept of judgment, is essentially 'equivocal.' On the one hand, the latter denotes the act of judging, a conscious experience which Husserl understands in terms of *"Fürwahrhaltungen,"* and which belongs to a class of concepts whose study pertains to psychology.[22] On the other hand, 'judgment' refers to objective propositions, *'Sätze,'* ideal forms that belong to pure logic. This act/content distinction bears on both the question of the theoretical foundations of logic and the conditions of the possibility of a theory in general, with respect to which we may adopt one of two opposed standpoints. Either we adopt an objective or logical standpoint whereby these conditions are grounded in the 'content' of the theory as such, that is, in the laws, principles, axioms, etc., or we accept the noetic standpoint whereby these conditions are grounded in the knowing subject. The significance Husserl ascribes to the noetic standpoint clearly manifests itself in his critique of sceptical relativism and in his remarks on the theory of knowledge in the last chapter of *Prolegomena*. In it, Husserl defends the thesis according to which the rational justification of a theory and thus the theoretical foundation of logic, rests on the evidence of the knowing subject (or apodictic and evident knowledge). One thus understands why this theory of knowledge plays such an important role in *Prolegomena*.

Hence it is important to situate this theory of knowledge in its relation to the *mathesis*. Husserl approaches this question through the idea of a division of labour with respect to the realization of the tasks of pure logic: the construction and edification of the *mathesis universalis* is taken over by the mathematician whom Husserl compares to a resourceful mechanic and technician. The *mathesis* is distinct from philosophical logic, which is to elucidate primitive logical concepts, their elementary forms of connections, and the laws that govern them (this task is accomplished in *Investigations* I, III, and IV). In contrast to Frege, Husserl understands the task of philosophy in *Prolegomena* not in terms of logical or linguistic analysis, but in terms of phenomenological analysis. However, this conceptual or eidetic analysis, inasmuch as it seeks "to achieve insight in regard to the sense and essence of his achievements as regards method and manner," belongs to the theory of knowledge.[23] In fact, the theory of knowledge serves as a "philosophical complement" to pure *mathesis* and Husserl assigns it a crucial role in *Prolegomena*. Its task is "to grasp perspicuously, from an objectively ideal standpoint, in what the possibility of perspicuous knowledge of the real consists, the possibility of science and of knowledge in general."[24] Since *Prolegomena* is rather expeditious as to the meaning of this theory of knowledge, I suggest a new distinction within this theory of knowledge. On the one hand is phenomenological analysis, which describes primitive logical concepts. On the other is the philosophical ideal of justification inherited from Descartes. This distinction is important for the purpose of showing the relative independence of phenomenology from the ideal of complete evidence and truth in itself, an ideal which Husserl will give up at the end of the 1920s. Apparently, the two merge in the very idea of justification and the following passage from *Prolegomena* would seem to summarize Husserl's position: "Ultimately, there-fore, all genuine, and, in particular, all scientific knowledge, rests on inner evidence: as far as such evidence extends, the concept of evidence extends also."[25] However, if evidence is understood in a properly phenomenological sense, that is, as consciousness of an original givenness *Gegebenheit* or as *Selbstgegebenheit*, we may say that phenomenology is subordinate to this philosophical ideal, but that it is nevertheless perfectly independent of it.[26]

Husserl ascribes to the phenomenology of *Logical Investiga-tions* a double task. It must analyse and describe, in essential generality, conscious experiences "treated as real classes of real events in the natural context of zoological reality, receive a scien-tific probing at the hands of empirical psychology."[27] The second task concerns the analysis and description of fundamental con-cepts and ideal logical laws. Our interest is focused on the latter task. The phenomenologist's first concern is to elucidate scientifi-cally the primitive concepts that make "the interconnected web of knowledge as seen objectively, and particularly the web of theory."[28] More precisely, he has to elucidate the elementary patterns (conjunctive, disjunctive, or hypothetical) according to which propositions may be connected to form new propositions as well as the concepts of object, state of affairs, number, relation, and the meaning categories, etc. This explanatory *Aufklärung* undertaking, which is essential to the theory of knowledge, does not proceed from explications *Erklärung* as is the case in the empirical sciences. Husserl, the student of Brentano, espouses the idea that the explication of a phenomenon in psychology, as well as the use of a concept in logic, depends directly on the description which is made of it. This description that proposes to account for the intelligibility of logical or psychological concepts is nothing else than *Aufklärung*:

> This 'clearing up' *Aufklärung* takes place in the framework of knowl-edge, a phenomenology oriented, as we saw, to the essential struc-tures of pure experiences and to the structures of sense *Sinnbestände* that belong to these.[29]

This 'clearing up' thus consists of investigating the origins of these concepts. But we now know that this investigation of the origins should not be understood as, for instance, the psychological genesis of the conceptual representations, but rather as the investigation of the *Einsicht in das Wesen* of these concepts. These preliminary tasks are taken up in the six *Logical Investigations* and represent indeed the main difficulties of the work.

4.

We now see that a Fregean reading of the phenomenology of *Logical Investigations* is justified as long as we confine ourselves to meaning and to what we called the logical motive, i.e. to the analysis of the constituents of the act in which ideal meaning and reference to an object consist. But what about the psychological motive – by which we should now understand to be the phenomenological motive – and the subjective dimension of the act to which the analysis of fundamental logical concepts takes us and which, as mentioned above, makes up the greatest part of the work? Frege rejected the division labour between mathematics and philosophy and maintained towards phenomenology the attitude he held in his review of Husserl's first book. One reason for this rejection is mentioned in Frege's often-quoted passage that speaks of a gap dividing mathematicians from psychologist logicians and of the latter's marked interest in the sense of acts and of representation in general.[30] This reason concerns the very meaning of the analysis of logical concept in terms of definition. Frege's remark echoes the critique Husserl makes of him in *Philosophie der Arithmetik* regarding the analysis of number. This apparently technical and isolated disagreement on the analysis of primitive logical concepts gains in philosophical significance when one considers that what is ultimately at stake in this dispute is no less than phenomenology's claim to situate the origins of these concepts in experience. In other words, what is at stake is the philosophical relevance of the psychological motive which first appears in *Philosophy of Arithmetic* and continues throughout Husserl's subsequent works. The present section will identify the object of the dispute. The two subsequent sections will reconstruct the essence of Frege's position on this question at difference places in his work.

First, let us recall that the chief aim of *Philosophy of Arithmetic* was to ground arithmetic in the concept of cardinal number and that the analysis of the concept, as Husserl already had mentioned in his 1887 dissertation, intrinsically belongs to psychology.[31] At the very beginning of the chapter on the formation of the concept of quantity, Husserl opposes psychological analyses to analysis in terms of logical definitions and clearly points out that:

> wir es nicht auf eine *Definition* des Begriffes Vielheit, sondern auf eine psychologische Charakteristik der Phänomene abgesehen haben, auf welchen die Abstraktion dieses Begriffes beruht.[32]

It is this very opposition between logical analysis and psychological analysis which leads him, in one of the appendices to chapter six, to assess Frege's position in *Foundations of Arithmetic*. According to Husserl, the logical foundations Frege is seeking for arithmetic in this work amounts to conceiving it as "eine Folge formaler Definitionen, aus welchen dis sämylichen Lehrsätze dieser Wissenschaft rein syllogistisch gefolgert werden können."[33] Whence the passage in *Foundations of Arithmetic* that Husserl quotes and where Frege claims that mathematics "jede Beihilfe von der Psychologie verbitten muss" and turns itself toward logic.

The comments that directly follow this passage are particularly interesting. They concern Husserl's objection to Frege's conception of logical analysis as definition. "Definieren kann man doch nur das logisch Zusammengesetzte. Sobald wir auf die letzten, elementaren Begriffe stoßen, hat alles Definieren ein Ende."[34] In dealing with elementary concepts, examples of which are all the concepts enumerated in the previous section as well as those of 'equality,' 'whole,' 'parts,' 'quality,' 'place,' 'time,' etc., it is recognized that these are not logically and formally definable concepts. When faced with simple and primitive concepts, it is often helpful to resort to language and use 'periphrasis': our objective each time is to provide a description of the concept which makes the understanding of its meaning possible. In such cases, one can also, according to Husserl "die konkreten Phänomene aufwei[sen], aus oder an denen sie abstrahiert sind, und die Art dieses Abstraktionsvorganges klarleg[en]."[35] Whether this process be understood from psychology, as is the case in this work, or in the sense of a doctrine of essence, as the passages of *Logical Investigations* quoted above imply, analysis is in both cases directed towards the phenomena from which these concepts are abstracted.

This is precisely what Frege, who replies directly to Husserl's objection that primitive mathematical terms are indefinable, disputes in his review. This reply appears to be unsatisfactory since it is content with bypassing the objection and with insisting instead on the disparity which opposes the psychologist's interests, who is concerned with the meaning of concepts which he moreover confuses with representations, and those of the logician who is concerned with the extension of concepts.[36] Nevertheless, in many texts which will be examined in the following section, Frege acknowledges the validity of this objection and grants it the highest significance, as is evident in his December 1899 letter to Hilbert. This letter is particularly interesting since Husserl comments on it in an appendix to his text *Das Imaginär in der Mathematik*.[37] Husserl starts with Frege's division of mathematical propositions into definitions and other propositions. In particular, he insists on the distinction made between definitions and what Frege calls *Erläuterungen*, that is, propositions or commentaries aiming at explaining or elucidating the meaning of the elementary concepts by means of which we define *definiens*. Husserl then quotes a long passage of this correspondence in which Frege explains in which context such definitions are required and how to use them. Frege writes about these *Erläuterungen*:

Auch sie enthalten also etwas, dessen Bedeutung wenigsten nicht als vollständig und unzweifelhaft bekannt vorausgesetzt werden kann, weil es etwas in der Sprache des Lebens schwankend oder vieldeutig gebraucht wird. Wenn in einem solchen Falle die beizulegende Bedeutung logisch einfach ist, so kann man keine eigentliche Definition geben, sondern muss sich darauf beschränken, die im Sprachgebrauch vorkommenden, aber nicht gewollten Bedeutungen abzuwehren und auf die gewollte hinzuweisen, wobei man freilich immer auf ein entgegenkommendes erratendes Verständnis rechnen muss. Solche Erläuterungssätze können bei den Beweisen nicht gleich den Definition gebraucht werden, weil ihnen die dazu nötige Genauigkeit fehlt, weshalb ich sie, wie gesagt, in den Vorhof verweisen möchte.[38]

Frege adds that these clarifications or explanatory propositions do not, properly speaking, belong to mathematics and this is why he consigns them "in the vestibule of mathematics" and relegates them to the role of "propedeutic." Husserl then reproduces some excerpts from this correspondence without, however, commenting on them. Nevertheless, Husserl's selection of passages and his introduction of the theory of definite manifolds (which he introduces in *Das Imaginär in der Mathematik* to which the commentary on the correspondence is annexed) testify to his inclination for Hilbert's axiomatic.

5.

Apparently, both the chief goal Frege ascribed to philosophy, that is, the investigation of the laws of truth, and the method he developed to achieve this goal – logical analysis – significantly increases the distance from phenomenology, even that of *Logical Investigations*. As previously suggested, any investigation directed towards the subjective dimension of acts and experiences would in all probability appear suspicious. Frege would readily recognize that *Logical Investigations* cannot be charged with the psychologism that he attributed to *Philosophy of Arithmetic*, namely, the confusion of the number with its subjective representation.

His distrust of phenomenology was motivated by the universal status he ascribed to logic and by the restrictions imposed on him by logical analysis. This is at least what emerges from his remarks on negation in his 1919 article *"Die Verneinung."*[39] As we know, negation features in two of the fundamental principles of Fregean logic, namely in that of the excluded middle and in the principle of non-contradiction. Negation and implication are the primitives of Frege's propositional calculus, from which all other operators may be derived – although this special status is not logically justified. It is unimportant from a strictly logical point of view, that such and such connector be the *definiens* and the other the *definiendum*. The philosophical question which is of interest to Frege in *"Die Verneinung"* relates to the problem of finding an interpretation of this concept which would concur with his conception of the thought and of the proposition through which it is expressed.[40] Hence, let us consider negation and other primitive logical concepts of Frege's logic.

Beginning with the proposition, one must mention one of the chief contributions of the *Begriffschrift*, namely, the substitution of the traditional conception of proposition (subject/copula/predicate) by an analysis in terms of function and argument. Thus understood, the elements of a proposition are not, contrary to what was assumed traditionally, the result of the synthesis operated on two substantives (for which the subject and the predicate would stand) but of a process of completion *Ergänzen* between a saturating part of the proposition (the object) and its unsaturated part (the function). This process of completion (of an element) would account for all cases of composition of parts into a whole: "And it is natural to suppose that, for logic in general, combination into a whole always comes about by the saturation of something unsaturated."[41] One may illustrate this need of completion by the expression "the negation of" or by the German genitive: "Die Verneinung des Satzes des Gedanken A." In the propositions "a is not equal to b" and "a is equal to b," for instance, one will simply say that the thought that "a is not equal to b" is the negation of "a is equal to b."

At any rate, this does not imply that there are two types of judgments, one negative the other affirmative, as is the case in Aristotle's logic for instance. We should not understand Frege's idea that "what I have just been designating as the polar opposite of judging I will now regard as a second way of judging" in the sense of there being a second type of *judgment*, but in the sense in which any assertive proposition, be it subject to negation or not, is an *affirmation*.[42] Hence, in order to reply negatively to a question one will say "It is false that," assuming that these words have both assertive force and an affirmative character.[43] In Peter Geach's words, to negate is to assert or affirm the negation of a proposition.

This distinction is clearly illustrated in the notation of the *Begriffsschrift* where the symbol of negation always follows that of assertion (| -). The notation for the negation of |-P is thus: | ⊥ P where the vertical stroke in the centre indicates that the negation pertains to the content of the assertion (the horizontal stroke '-') and that it does not modify the assertive force which is indicated by the first vertical stroke ' | '.[44] The letter P stands for a proposition which is understood as the expression of a *Gedanke* or thought, properly speaking, what can be said to be either true or false. One needs to distinguish propositions, thus understood, from judgments as well as from the act of thinking. Judgments are what allow us to go from the meaning or thought to the *Bedeutung* (reference), in this case, to the truth value of a proposition. It is a mere 'judgment of recognition' since it consists precisely in recognizing such or such proposition as true or false.[45] In the *Begriffsschrift*, the judgment is designated through the symbol ' | -' and the occurrence of this symbol indicates that the sign which follows it is used in an assertive manner or with *Behauptungskraft*.[46] On the other hand, the act of thinking corresponds to what Frege calls "to grasp a thought" *Faßen eines Gedanken* and to think, according to him, means nothing else than to grasp a thought.[47] But there is a difference between grasping and representing a thought. Frege concedes that something in consciousness must be directed towards *hinzielen* this thought, but this something pertains to subjective representation and must be distinguished from the thought as such which does not require any bearer. Frege compares grasping of thought to holding an object in one's hand:

> What I hold in my hand can certainly be regarded as the content of my hand; but all the same it is the content of my hand in quite another and a more extraneous way than are the bones and muscles of which the hand consists or again the tensions these undergo.[48]

Of course, one must concede that the way in which the content of the hand is articulated is different from the way in which the hand itself is articulated. Must we not, however, presuppose that thought (in the subjective sense) is constituted in a certain manner in order to be able to grasp (objective) thoughts? Must we not, using an Aristotelian metaphor most opportune in a context which builds on the manipulation of *pragmata*, conceive thought or *noema* not as the object which is grasped, even less as that through which we grasp, but as a structure which, like the joints of the hand, is necessary to its being grasped? One seeks, in vain, in Frege's work, an anwser to this question. The meaning of the idea of grasping a thought, which is an essential feature of his theory of judgment and which points to Husserl's doctrine of acts, is left undetermined. I would like to suggest that we are not dealing here with mere negligence on Frege's part, nor with a prejudice against psychology, but with one of the consequences of the application of one of the leading principles of his philosophy.

6.

Why exactly Frege's theory lacks an analysis of the notion of grasping a thought – as it also lacks an analysis of the notion of negation which, as a simple element of the system, is indefinable – may be understood in the light of his account of a particular mistake of the philosophical tradition. It concerns, Frege explains in "The Negation," the need we have to define the concepts we use:

> It is certainly praiseworthy to try to make clear to oneself as far as possible the sense one associates with a word. But here we must not forget that not everything can be defined. If we insist at any price on defining what is essentially undefinable, we really fasten upon inessential accessories, and thus start the inquiry on a wrong track at the very outset.[49]

This is the case of the notion of judgment to which Frege alludes in this passage, but it holds *a fortiori* for all fundamental philosophical concepts in his theory, namely: the truth of objects,[50] concepts,[51] and functions.[52] In each case we are dealing with logically simple elements and consequently with concepts which cannot be logically defined or analysed. The mistake to which Frege alludes in this passage clearly arises when we attempt to define the term 'concept,' which is of a predicative nature or in logical terms a monadic function. To define it, we must transform it into the object of a new predication. But this transformation violates *Grundlagen*'s third principle, which states that concepts and objects must be separated. Whence paradoxes such as "the concept horse is not a concept" or "the function f(x) is not a function," paradoxes which may be ascribed to the objectifying nature of language. For, Frege explains, we use a nominal turn of phrase and we intend a concept. This is why we cannot talk about the sense of a non-saturated expression without turning it into a proper name and, in the case of concept, without obliterating the predicative nature of this expression.[53] Faced with this apparent limitation of logical analysis and with the absence of definition for simple logical elements, Wittgenstein employed the well-known distinction between saying and showing, a distinction which analytical philosophy's orthodoxy has termed mystifying.[54] But we could show, as Wittgenstein did in his *Tractatus*, that this distinction significantly affects the distinction between *Sinn* and *Bedeutung*. If the logically simple elements cannot be defined, it is because a definition sets out what the *reference* of expression should be, but not its *sense*. However, in saying what the reference should be and in choosing a particular way of doing it, it shows what the sense should be.[55]

But must we not assume, as Wittgenstein also suggests in the passage in *Tractatus* we have drawn out, a minimal and perhaps implicit knowledge of the meaning of, for instance, the notions of negation, concept, object, or function, knowledge which only definitions seem to be able to provide? Frege agrees and asserts that in the absence of a definition of these simple elements, we must replace them, temporarily of course, by an explanatory commentary *Erläuterung*.[56] This *Erläuterung*, as we recall, does not, properly speaking, pertain to logic, and its purpose is to show by means of examples taken from ordinary language what cannot be said from within the system, that is, the meaning of terms which must be known even before we use them. For logic, according to Frege, is *lingua characteristica* and not *calculus ratiocinator* as Boole and Schröder believed, and as such, it must be learned. Explanatory commentaries, which are not necessary for the 'isolated scholar' thus fulfill a practical aim which consists in granting to the scientific community cognitive access to these meanings. It uses the figurative character of expression, as we have done when elucidating the idea of grasping a thought using the metaphor of the hand. But resorting to metaphors becomes problematic when we know Frege's distrust of ordinary language. I will not insist on this point although I believe that it is an important one in the actual context. One needs merely mention that explanatory commentary stops at the very moment when we grasp the meaning of these notions and that we must be satisfied with it.

The question is, of course, why should we bind the analysis of these concepts to the metaphors and periphrases of an explanatory commentary? Why not entrust the task to phenomenological investigations, for instance? A straightforward answer to this question is to be found in one of the dogmas of classical analytical philosophy and more precisely in logical positivism. It consists in subordinating all philosophical questions to logic. Although this postulate is rarely made explicit by Frege, it imposes major constraint on all of his philosophical moves, both at the ontological and at the methodological level. At the ontological level, Frege's universe is essentially composed of concepts and functions and this universe is not arbitrary since it entirely exhausts our ontology. At the methodological level, the universal character of logic implies, as van Heijenoort has shown, that nothing can be said from outside the system.[57] This might explain why Frege does not ask metalogical questions (consistence, completeness, independence of axioms, etc.) and that he would reject any solution to this problem which would go in the sense of "semantic ascension" (substitution of the language about objects by a language about expressions, or of the material language by a formal metalanguage). For whatever the language used and whatever the level of the hierarchy, it necessarily presupposes the distinction between concepts and objects (saturated/non saturated) which, as we have said above, is the touchstone of the system.

FINAL REMARKS

Frege's standpoint would reject the phenomenological call to return to the things themselves in terms of evidence. Was it not precisely the sense of Frege's criticism of Husserl and the psychologists in his 1894 review where he claims that the difference between logic and psychology supervenes on the difference between, on the one hand, the logician's marked interest for *Bedeutung*, i.e. truth, and on the other hand, the psychologist's poet-like interest for the *Sinn* or meaning? If this is truly the case, antipsychologism is a necessary but not a sufficient condition to escape Frege's critique. Yet, is this critique really legitimate? Does not denying to psychology the status of first philosophy and entrusting this role to logic expose one to objections which contemporary philosophy recognizes as largely compelling? Let us keep in mind that those are precisely the presuppositions which Quine has condemned in order to rehabilitate the philosophical status of the field of study Husserl seeks to open with his phenomenology in *Logical Investigations*. Quine has indeed shown that Frege's, Wittgenstein's, and the Logical Positivists' idea that logical analysis consists in an *a priori* exercise entirely separated from sensible experience presupposes the analytic/synthetic distinction. Yet, as Quine has shown in "Two Dogmas of Empiricism," the latter is not well-founded.[58] In particular, Quine disputes the dichotomy between statements that pertain to logic (analytic) and those that pertain to empirical sciences (synthetic), and proposes a form of semantic holism according to which the difference between these two types of statements is one of degree. One is aware of the influence this critique has exerted on contemporary philosophy and, as we have already noticed, it has largely contributed to rehabilitate the philosophy of mind. But, in Quine, this rehabilitation goes hand in hand with a return in strength of philosophical naturalism, which has a family resemblance to the psychologism Husserl and Frege criticized. Quine believes that epistemological questions as well as all those questions that traditionally pertain to philosophy can be replaced by questions that pertain to psychology and, in particular, to stimuli/response behaviourist psychology. But this radical naturalism also has its share of problems, namely with the *qualia* or what is also called phenomenal experience. The descriptive framework which it borrows either from neurology, biology, or from behaviourist psychology does not, according to many, have at its disposal the necessary resources for the description and explanation of this fundamental dimension of human experience. In this respect, phenomenology, whose philosophical focus is based precisely on this dimension, might be of great assistance. And for this purpose, phenomenology has developed a conceptual framework whose purpose might be compared to Frege's use of *Erläuterungen* in the elucidation of primitive (logical) notions.[59]

Notes

1   Edmund Husserl, *Aufsätze und Vorträge*, Husserliana Band XXV (den Haag: Martinus Nijhoff, 1970), 298.

2   Husserl, *Aufsätze und Vorträge*, 298.

3   See D. Fisette, *Lecture frégéenne de la phénoménologie* (Paris: Éclat, 1994).

4   At any rate, it is to Frege that Husserl refers at §15 of the first *Logical Investigation*, when he introduces the distinction.

5   Useful information on what is at stake in these debates can be found in Martin Kusch, *Psychologism* (London: Routledge, 1995). I have examined the difference between normative antipsychologism and antipsychologism that resorts to the ideality of logical laws. See D. Fisette, « L'antipsychologisme de la phénoménologie et la psychologie », in *Frege: logique et philosophie*, ed. M. Marion (Paris: l'Harmattan, 1998).

6   Michael Dummett, *Origins of Analytical Philosophy* (Cambridge: Harvard University Press, 1993).

7   Pierre Poirier and Denis Fisette, eds., *Problèmes de conscience* (Paris: l'Harmattan, 2001).

8   Husserl, *Logische Untersuchungen: Unterschungen zur Phänomenologie und Theorie der Erkenntniss*, Husserliana Band XIX/1, tr. J.N. Findlay, *Logical Investigations*, 2 vols. (London: Routledge and Kegan Paul, 1970), 7.

9   On the problem of the relation between intentional psychology and phenomenology in Husserl as well as on Husserl's conception of the task of the former, see his 1925 lectures, *Phänomenologische Psychologie:Vorlesungen Sommersemester*, Husserliana, Band IX (1962).

10 Gottlob Frege, *Collected Papers* (London: Blackwell, 1984), 198.

11 Husserl, "Entwurf einer 'Vorrede' zu den 'Logischen Untersuchungen,'" *Tijdschrift voor Filosophie* 1 (1939): 112–13.

12 Husserl, *Logische Untersuchungen: Prolegomena zur reinen Logik*, Husserliana Band XVIII, 7 (tr. Findlay, 12).

13 Husserl, "Review of Melchior Palágyi's "Der Streit der Psychologisten und Formalisten in der modernen Logik," *Zeitschrift für Psychologie und Physiologie der Sinnesorgane*, (1903), 31: 287–294. English translation: "Early Writings in the Philosophy of Logic and Mathematics," in *Edmund Husserl Collected Works*, tr. and ed. Dallas Willard (Boston: Kluwer, 1994), 5: 197–206.

14 Husserl, *Logische Untersuchungen: Unterschungen zur Phänomenologie und Theorie der Erkenntniss*, 23 (tr. Findlay, 28).

15 In *Prolegomena*, the use of "descriptive science" refers clearly to what von Kries called ontological sciences (geography, natural sciences, anatomy) and are opposed to properly nomological sciences which Husserl characterizes as explicative sciences. *Prolegomena zur reinen Logik*, 237 (tr. Findlay, 240).

16 Husserl, *Logische Untersuchungen: Elemente einer phänomenologischen Aufklärung der Erkenntnis*, Husserliana, Band XIX, 756 (tr. Findlay, 760).

17 Husserl, *Logische Untersuchungen: Unterschungen zur Phänomenologie und Theorie der Erkenntniss*. 23 (tr. Findlay, 28).

18 See Husserl's 1901 lecture "*Das Imaginär in der Mathematik.*" Husserliana XII.

19 Husserl, *Logische Untersuchungen: Prolegomena zur reinen Logik*, 23 (tr. Findlay, 28).

20 Husserl, *Logische Untersuchungen: Prolegomena zur reinen Logik*, 162 (tr. Findlay, 169).

21 Husserl, *Logische Untersuchungen: Prolegomena zur reinen Logik*, 176 (tr. Findlay, 181).

22 Husserl, *Logische Untersuchungen: Prolegomena zur reinen Logik*, 178 (tr. Findlay, 182).

23 Husserl, *Logische Untersuchungen: Prolegomena zur reinen Logik*, 255 (tr. Findlay, 245).

24 Husserl, *Logische Untersuchungen: Prolegomena zur reinen Logik*, 209 (tr. Findlay, 207).

25 Husserl, *Logische Untersuchungen: Prolegomena zur reinen Logik*, 25 (tr. Findlay, 61).

26 Husserl, *Logische Untersuchungen: Prolegomena zur reinen Logik*, 193 (tr. Findlay, 198).

27 Husserl, *Logische Untersuchungen: Unterschungen zur Phänomenologie und Theorie der Erkenntniss*, 7 (tr. Findlay, 249).

28 Husserl, *Logische Untersuchungen: Prolegomena zur reinen Logik*, 244 (tr. Findlay, 236).

29 Husserl, *Logische Untersuchungen: Unterschungen zur Phänomenologie und Theorie der Erkenntniss*, 27 (tr. Findlay, 265).

30 Frege, *Collected Papers*, 200.

31 Husserl, *Philosophie der Arithmetik*, Husserliana Band XII, 1970, traduction française, J. English, *Philosophie de l'arithmétique* (Paris: PUF, 1972).

32 Husserl, *Philosophie der Arithmetik*, 20 (English translation, 25).

33 Husserl, *Philosophie der Arithmetik*, 118 (English translation, 145).

34 Husserl, *Philosophie der Arithmetik*, 119 (English translation, 146).

35 Husserl, *Philosophie der Arithmetik*, 119 (English translation, 146).

36   Frege replies to this objection in his review, *Collected Papers*, 200. He takes as example two definitions of a cone and asserts that, for a mathematician, the choice of one of these definitions is motivated by "reasons of convenience."

37   Husserl, "*Das Imaginär in der Mathematik.*"

38   Husserl, "*Das Imaginär in der Mathematik,*" 9.

39   This article is the second of a series of three texts titled *Logische Untersuchungen*, the first being "Thoughts" and the third "Compound Thoughts." The significance Frege ascribes to this investigation emerges from a passage of his "Kurze Übersicht über meiner logischen Lehren," *Collected Papers*, 214, where he claims that one of his chief contributions to logic consists precisely in having succeeded in liberating the concept of negation from the traditional conception of judgment, the Kantian in particular, which according to him was reducing it to an activity merely parallel to assertion. There are other reasons for favoring the question of negation in the present context. Apart from the fact that the elucidation of this concept represents one of the tasks assigned to phenomenology by Husserl in *Logical Investigations*, the interpretation of negation is an important topic in the few letters Husserl and Frege exchanged in 1906. In particular, Frege's letter of 1 November 1906, in which he confirms the receipt of Husserl's five reviews which Husserl published in 1904 in *Archiv für systematische Philosophie* No. 10, of works that appeared in Germany in the years 1895–1899. The fifth review (of Anton Marty's "Über subjektlose Sätze und *das Verhaltnis der Grammatik zur Logik und Psychologie*") discusses negation. Frege's letter shows that he indeed agreed with Husserl's critique of Marty.

40   In fact, a good deal of this article is an assessment of the implication of the interpretation for the traditional conception of propositions and judgments where negation is understood as an act opposed and parallel to that of judgment. "Even Kant does it." Frege, *Collected Papers*, 380. According to this conception, "the judging subject sets up the connection or order of the parts." *Collected Papers*, 381. Moreover, the negation of a thought would be the "dissolution of the thought into its component parts." *Collected Papers*, 377. One argument against this conception of negation emphasizes the fact that there are no criteria, grammatical or of another nature, which would make possible a clear distinction between a negative and a positive judgment. Hence, in "Concepts and Objects," *Collected Papers*, 187, Frege resorts to negation in order to demonstrate that the terms "all," "none," and "some" which are considered to be grammatical subjects, always appear in

connection with a conceptual term although they relate to the whole proposition (in Frege's sense). But we may also show that a singular term cannot bear negation. That is what the negation of the following proposition shows: 1) "All mammals live on Earth." If the words "All mammals" were the subject of the predicate "live on Earth," in order to negate the whole, we would have to negate the predicate "do not live on Earth." But, it is clear that the negation or the thought opposite to 1) is not: 2) "All mammals do not live on Earth," rather: 3) "No mammals live on Earth." It is thus advisable to put the negation before "all." Whence the thesis according to which "all" belongs logically to the predicate to which the negation applies and never to the grammatical subject. Nevertheless, the occurrence of "do not" in the predicate should not make us forget that negation, which is a component of meaning, relates to the whole proposition.

41 Frege, *Collected Papers*, 390.

42 Frege, *Collected Papers*, 383.

43 As Frege writes in *"Logik"*: "Die Behauptung liegt dabei wie sonst in der Form des Indikativ und ist nicht notwendig mit dem Worte nicht verbunden" *Collected Papers*, 70.

44 As Frege points out in this text, the primary upshot of the rejection of the traditional conception is of an economical nature. By dissociating negation from judgment and turning it into an element of meaning, there remains only affirmation and a term designating negation. This obviates the need for an extra inference principle. If we permitted two types of judgment, affirmative and negative, we would also require two types of inference:

$$\neg P \rightarrow \neg Q \quad \text{and} \quad S \rightarrow \neg Q$$
$$\underline{\neg P} \qquad\qquad\qquad \underline{S}$$
$$\neg Q \qquad\qquad\qquad \neg Q$$

But it is clear that, in both cases, we are dealing with *modus ponens*. On this question, see *Logical Investigations* 380 and 384–5.

45 On this theory of judgment and for useful remarks on the notation of the *Begriffsschrift* see D. Bell, *Frege's Theory of Judgement* (Oxford: Clarendon Press, 1979), 83ff.

46 One must however note that the scope of negation extends beyond the domain of assertive propositions to interrogative propositions, optative propositions, imperative propositions, etc. Hence, the scope of assertive force is much narrower than that of negation since it does not apply to questions, to a nominal phrase "that P" to "¬P" or to any other proposition that is involved in a conditional or a disjunction. The latter can be ex-

plained by the fact that assertive force cannot be ascribed to a false statement, but a conditional or disjunctive statement may be true although one
of its members is false. It is otherwise with conjunction, where all members
must be true in order for the whole thought to be true. Any statement
involving a connector loses its assertive force since its meaning becomes
indirect and must be completed. *Collected Papers*, 404.

47 Frege, *Collected Papers*, 368.

48 Frege, *Collected Papers*, 368.

49 Frege, *Collected Papers*, 381.

50 In "Funktion und Begriff," *Collected Papers*, 18, Frege claims that once we
    have conceded that objects are arguments and values of functions, the
    question arises as to what we should understand by 'object.' He answers:
    "Eine schulgemässe Definition halte ich für unmöglich, weil wir hier etwas
    haben, was wegen seiner Einfachheit eine logische Zerlegung nicht zulässt.
    Es ist nur möglich, auf das hinzudeuten, was gemeint ist. Hier kann nur
    kurz gesagt werden: Gegenstand ist alles, was nicht Funktion ist, dessen
    Ausdruck also keine leere Stelle mit sich führt."

51 The same remark is to be found in "*Über Begriff und Gegenstand*" with
    respect to the definition of 'concept.' "Was einfach ist, kann nicht zerlegt
    werden, und was logisch einfach ist, kann nicht eigentlich definiert werden,"
    Collected Papers, 193.

52 In "Was ist eine Funktion?" we find the same comment as well as in many
    other passages of "*Logik in der Mathematik*" such as the following: "Durch
    eine Definition ist es nicht möglich anzugeben, was eine Funktion ist, weil
    es sich hier um etwas Einfaches und Unzerlegbares handelt. Es ist nur
    möglich, auf das Gemeinte hinzuführen, und es durch Anknüpfung an
    Bekanntes deutlicher zu machen. An die Stelle einer Definition muß eine
    Erläuterung treten, die freilich auf ein entgegenkommendes Verständnis
    rechnen muß." *Collected Papers*, 142.

53 And if this were the case, we might want to ask whether explanatory
    commentary leads anywhere since it brings us back to ordinary language
    whose inexactness we know. It is interesting to note that, according to
    Frege, the major problem with ordinary language is its tendency to pro-
    duce proper names to which no referent corresponds. If these expressions
    are essential to poetry – let us not forget that Frege understands herewith
    all the sciences, including sciences of the mind, which show an interest for
    meaning – they are also at the source of some paradoxes or logical antino-
    mies. Indeed, Frege believes that Russell's antinomy in the set theory is an
    upshot of the attempt to provide a logical foundation for numbers, that is,

of conceiving them as sets. Frege, *Nachgelassene Schriften* (Hamburg: Meiner, 1969), 1: 288–289. In saying for instance "the extension of concept a" or "the concept of fixed star," neither expression has an object. The definite article which is the mark of a logical proper name seems to indicate that this expression denotes a concept while it denotes nothing. Frege will thus say that, in this case, we are dealing with a pseudo-proper name.

54   Dummett, *The Interpretation of Frege's Philosophy* (Cambridge: Harvard University Press, 1981). Dummet holds this point of view in the chapter on definitions.

55   See Dummet's article "Frege and Wittgenstein," in *Perspectives on the Philosophy of Wittgenstein*, ed. E. Block (Cambridge: Harvard University Press, 1981), 33.

56   On that question, see Peter Geach's interesting paper "Saying and Showing in Frege and Wittgenstein," *Acta Philosophica Fennica* 28 (1976): 54–70.

57   Van Heijenoort, "Logic as Language and Logic as Calculus," *Boston Studies in Philosophy of Science* (1967), 3: 3.

58   W.V.O. Quine, "Two Dogmas of Empiricism," in *From a Logical Point of View* (Cambridge: Harvard University Press, 1953), 20–46.

59   On Quine's critique of the logicist tradition and on the significance of Quine's naturalism for the expansion of cognitive sciences and the philosophy of mind since the end of the 1950s, I refer the reader to Pierre Poirier and Denis Fisette, *La philosophie de l'esprit : état des lieux* (Paris: Vrin, 2000).

PART II

PHENOMENOLOGY, MATHEMATICS, AND PHYSICS

CHAPTER FOUR

Ulrich Majer

HUSSERL AND HILBERT ON GEOMETRY

1. INTRODUCTION

Anyone who attempts to compare Husserl's and Hilbert's approach to geometry faces an almost insurmountable difficulty. Whereas Hilbert, over a period of more than ten years, worked out a systematic and detailed presentation of geometry which was published in his book Grundlagen der Geometrie, there is nothing comparable in Husserl's work.[1] All that we find in Husserl's Nachlaß[2] is a blueprint for a book on geometry, some scattered remarks about the epistemological origin of our knowledge of space, two somewhat longer scripts (one on the history of geometry, the other on topological questions), and last but not least, some shorter notes regarding the works of other geometers.[3]

Fortunately, one of Husserl's pupils, Oskar Becker (professor of mathematics at Bonn), wrote an extensive essay, the "Phenomenological Foundation of Geometry and its Application in Physics."[4] I shall use this essay to construct a preliminary description of Husserl's position regarding the epistemological foundations of geometry. Because one cannot be certain that Becker's position is identical to Husserl's (rather than some interpretation), I shall revisit this preliminary description by investigating Husserl's remarks on what he calls the 'different approaches in geometry' towards the end of the nineteenth century and his eventual affiliation with one of them. Husserl does not embrace Poincaré's conventionalism, as Becker suggests and most Husserl experts believe; rather, I shall show that Husserl embraces a position very much like Hilbert's axiomatic point of view.

2. DIFFERENT APPROACHES TO GEOMETRY

The science of geometry can be distinguished according to different points of view. The most common is the methodological distinction between analytic and synthetic geometry. The first uses the concept of number, whereas the second avoids the concept of number altogether. Another distinction is that between different kinds of geometries, such as Euclidean and non-Euclidean geometry. Both distinctions can be combined freely. Bolayi, for example, developed non-Euclidean hyperbolic geometry in a synthetic fashion. But these distinctions are not my main concern. If I speak of different 'approaches' to geometry, the distinction that I propose cuts across these traditional distinctions relates to a development in geometry during the late nineteenth century. It is best characterized as the distinction between a strict axiomatic approach and a non-axiomatic (more or less) intuitive approach. Because the term 'axiomatic' has different meanings, I shall explain which sense I use here.

The paradigm example of an axiomatic approach is presented in Euclid's *Elements* (although it does not withstand modern critical standards). Euclid's axiomatic approach was forgotten, or pushed aside, during the seventeenth and eighteenth centuries, first by the rise of analytic geometry and then by the development of projective geometry. Consequently, it had to be reanimated against the domination of the projective school. The projective school's most advanced and progressive expression was Felix Klein's so called *"Erlanger Programm."* The explicit intention of this programme was to unify the different kinds of geometry such as hyperbolic, elliptic, and plane geometry under the leadership of projective geometry, and more importantly, its group-theoretic point of view. I will explain this later in more detail. For the moment it is sufficient to understand that this programme had no essential connection to Euclid's axiomatic approach. It proceeded independently of any serious axiomatic considerations, guided primarily by a special kind of intuition and the new mathematical idea of a group. It could, of course, be brought into an "axiomatic form," at least externally, by present-ing the principles of projective geometry and group-theory as axioms, from which the Euclidean and non-Euclidean geom-etries can be deduced as special cases. But such a 'formal' axi-omatization has very little to do with the spirit and intentions of the 'new' axiomatic approach, as it emerged towards the end of the nineteenth century. The 'new' axiomatic approach was devel-oped foremost and primarily by two men, to restrict the modern history of geometry to Germany, Moritz Pasch and David Hil-bert. Because this 'new' approach is often misrepresented by exaggerating its opposition to Euclid, I shall briefly summarize its main intentions and methods.

The primary goal of the new axiomatic approach was a clari-fication and intensification of geometrical proofs. To this end Pasch demanded that each and every proof in geometry must proceed strictly deductively, which means two things. First, a proof must articulate all assumptions, even the most obvious, from which a given sentence shall be deduced. ("No silent pre-suppositions" was the slogan of the day.) Second, a proof must not rely on pictures or any other form of sensible intuition; every presupposition (necessary for the deduction of a particular sen-tence) must be explicitly expressed as a valid *proposition*, i.e., either as an axiom or a previously deduced sentence. The means and methods by which Pasch hoped to achieve this goal were twofold: first, a rigorous logical *analysis* of the intuitive content of the geometrical sentences; second, a strict *formalization* of the deduction according to the grammatical rules of the German language. Pasch, obeying his own maxim, was able to recognize the most elementary and for his time, completely unknown axioms of geometry, the topological axioms of ordering.

Hilbert followed Pasch's footsteps, but elevated the axiomatic procedure to an unprecedented and hardly imaginable level of sophistication by eliminating some of the restrictions and curiosities which characterized Pasch's approach. Hilbert recognized that Pasch's deductive procedure was too restricted and had to be supplemented by model-theoretic means in order to prove for example the logical *independence* of the axiom of parallels from the other Euclidean axioms. In this context Hilbert developed what he himself called the 'axiomatic method,' namely, the deliberate variation of a system of axioms, at will so to speak, in order to judge which sentence (or axiom) depends on which, and by this means to recognize the minimal set of independent axioms.

This method is frequently misunderstood by equating it to the axiomatic approach in general, but it is only a part of the whole axiomatic procedure, and has to be supplemented by an inquiry into the fundamental propositions of a certain field of facts, for example the facts of geometry, and consequently the establishment of a corresponding system of axioms. Which of these facts are objective in the case of geometry is another delicate question, which I shall discuss later. For now it is important to note that the axiomatic approach as a whole consists of two parts. First and foremost is the establishment of a system of axioms which corresponds to a certain field of facts. Second, the application of the axiomatic method as a means of inquiry into the logical relations that hold among the propositions of an axiom system, their logical dependence and independence, their completeness, and last but not least, their relative consistency. I call these relations the *metalogical* relations of axioms and axiom systems in distinction from the usual logical relations used within sentences. We can say that Hilbert amended the classical axiomatic approach of Euclid and Pasch with a systematic inquiry of the metalogical relations among axioms and axiom systems. The principal means of this inquiry is the axiomatic method in connection with a certain type of model-theoretic consideration, which was fairly new during Hilbert's time, but have meanwhile become a standard method of proof. One must be clear that the model-theoretic method was – and to some extent still is – the most problematic part of Hilbert's axiomatic approach and caused him a lot of trouble. It is therefore no wonder that he eventually suspended it in favour of a new, more direct method of proof, his famous proof–theory.

I now return to discuss the distinction between the different approaches to geometry. As mentioned, the distinction cuts across the traditional distinctions such as analytic and synthetic, Euclidean and non-Euclidean geometries. It concerns our understanding of geometry on different *epistemological* grounds, and therefore attempts to establish geometry, or better, the multiplicity of geometries and their interrelations, in fundamentally different manners. The first approach is the group-theoretic approach, as expressed in Klein's "*Erlanger Programm.*" It tries to ground different geometries and their interrelations in a group-theoretic analysis of the motions, i.e. the rotations and translations of 'rigid bodies' in space. The second approach is the new axiomatic approach of Pasch and Hilbert. It tries to ground the different geometries and their conceptual relations in a rigorous analysis of the *metalogical* relations between the various axioms and axiom systems by means of the axiomatic method. But, of course, every analysis presupposes something to analyse, and so the question arises. What is this 'something' in the case of geometry? The interesting point is that Hilbert's answer is stated quite explicitly in the *Grundlagen* , although it has been ignored by most readers because they believed that it would be incompatible with his axiomatic approach. I will quote Hilbert's answer and return to the difference between the two approaches.

For its construction geometry requires, like arithmetic, only few and simple principles. These principles are called axioms. The establishment of the axioms of geometry and the investigation of their relationships is a problem that has been discussed since Euclid in many excellent essays of the mathematical literature. This problem is equivalent to the logical analysis of our spatial intuition.[5]

The last sentence is, of course, the crucial one. Although it may be rather unclear what precisely Hilbert had in mind by the term "spatial intuition," at least so much is clear at this point. He thinks geometry has a non-arbitrary content that is somehow 'given' to us. Moreover, the manner in which it is given is somehow connected to our spatial intuition. It is for this reason that Hilbert can maintain that the task of the mathematician – the establishment of the fundamental principles or axioms of geometry – is equivalent to the "logical analysis of our spatial intuition." Otherwise this assertion would make no sense at all.

Notice that it remains open as to whether this intuition is empirical (constituted by some special kind of experience) or *a priori* (in the temporal, transcendental, or still another sense of *a priori*). So far we have the presupposition that we have epistemic access to this kind of intuition and that we know (or at least could know) the principles of geometry by a logical analysis of spatial intuition. Presumably this kind of intuition is the same for all human beings, or more specifically, for all trained mathematicians.

But what is the main difference between the group-theoretic approach of Klein, Helmholtz, Lie, and philosophically most importantly, Poincaré, and the new axiomatic approach of Hilbert?[6] It should be clear from what I have said so far that the two approaches stood not so much in strict *opposition* to each other but in a kind of *competition*, a competition about the most appropriate and profound way to a reliable *foundation* for geometry. It is important to note that the difference is not that the group-theoretic approach bases geometry on a particular kind of intuition (the motion of rigid bodies in space) whereas the new axiomatic approach rejects intuition and establishes geometry as a purely formal mathematical discipline, whose agreement with the real world has to be judged exclusively on empirical grounds. This is a serious misreading of Hilbert's axiomatic approach,

launched by the Logical Empiricists to support their own epistemology. But it is patently wrong, as I have tried to show earlier.
Instead, with respect to the question of intuition (as a source of
geometrical knowledge) we find both approaches on the same
side of the dividing line and in strong opposition to Logical
Empiricism, which rejects any recursion to pure intuition as a
source of geometrical knowledge. This is an important first result, insofar as Husserl regards himself likewise in strong opposition to Logical Empiricism.

Now, since both approaches accept intuition as a source of
geometrical knowledge, what divides them? To answer this question we have to examine what *kind* of intuition each approach
accepts as grounding geometry. Here a remarkable difference
occurs, which will help identify Husserl's position regarding the
epistemological foundations of geometry. The core of the difference is the following.

The group-theoretic approach of HLP takes the 'motion of
rigid bodies' as fundamental for the determination of the *relational* structure of space, and consequently as fundamental for
geometry (taken as a general theory of spaces of different structures). In other words, only insofar as we have a distinct intuition
of the motions of rigid bodies in space, in particular their possible translations and rotations, can we articulate the space's
geometrical structure and therefore its corresponding geometry.

In Hilbert's view, however, this approach presupposes too
much as intuitively given, which can and must be logically
analysed. The contents of this 'too much' will be addressed later.
For now it is sufficient to note that for this reason Hilbert had to
rely on another, more elementary, kind of intuition for a proper
foundation of geometry. It is difficult to state positively the kind
or form of intuition this is. Hilbert calls it simply our 'spatial
intuition' in distinction to the intuition of time. But one *negative*
characterization is at least clear. This kind of intuition excludes
any temporal aspect or moment, and consequently does not rely
on any assumptions regarding the *differentiability* of the translation and rotation functions with respect to time, as the group-

theoretic approach obviously does. Such an assumption is not only *superfluous* for a reliable foundation of geometry, as Hilbert showed in an important paper in 1902, but may also turn out to be wrong (in light of quantum mechanics). The recognition that this assumption can be abandoned is in Hilbert's view crucial for a deeper understanding of geometry. This is, of course, only a first step. The final analysis presupposes a completely different, totally atemporal kind of intuition. The proper foundations of geometry are the timeless qualitative and quantitative relations between bodies as they appear in our spatial intuition; this excludes the motions of bodies and their alleged properties from the foundations of geometry.[7]

3. HUSSERL'S APPROACH TO GEOMETRY

Is Husserl closer to the group-theoretic approach of HLP or does he share Hilbert's axiomatic approach? Because it is difficult to answer this question unambiguously from Husserl's own writings, let us first consider Becker's presentation of the phenomenological approach. It is, however, beyond the scope of the present paper to recapitulate Becker's phenomenological foundation of geometry in all its aspects and subtle details. Considerations will focus on the most robust aspects of his presentation that are relevant for our main question.

According to Becker, geometry is a rational science like arithmetic and analysis, but unlike these purely *formal* sciences, geometry deals with a content, which is *contingent* but nonetheless given *a priori*. This "double aspect," as Becker calls the contingent yet *a priori* given content, is characteristic for geometry and determines its unique status among the sciences. According to this "double aspect" a phenomenological foundation of geometry has to fulfil two tasks. First, it has to elucidate the nature of the *a priori* given but contingent content of geometry and how it is constituted. This is the genetic or the cognitive-psychological part of the phenomenological approach. Second, to make geom-

etry a rational science, a phenomenological foundation must sur-
mount the 'contingent' character of its content. That is, it must
explain why the *a priori* given content has the particular structure
that it does, rather than one of the many other possible structures
that from a logical perspective are equally realizable. In other
words, the rational task of the phenomenological approach is to
explain why a certain kind of geometry is necessary for human
beings, even thought it is contingent from a logical perspective.
Both tasks are interrelated. The genetic part leads only to a vague
and 'indefinite' idea of space, which the rational part clarifies
and determines. Here a type of idealization occurs (*ideierende
Abstraktion*), which results in a particular 'limit–transition' (*Gren-
zübergang*) to the absolute exact and definite space of Euclidean
geometry. Immediately after presenting the phenomenological
approach Becker explains the mathematical meaning of these tasks.

> The first problem (the rational apprehension of the plain intuitive
> content) is, expressed in the usual mathematical terminology, noth-
> ing other than the continuum–problem; the second (the surmount-
> ing of the contingent character), however, is the problem of the
> relation of the Euclidean to the so-called non-Euclidean geometries.[8]

Of the many points discussed by Becker, only two are relevant to this discussion, for our concern, namely 1) the genetic
constitution of space, in particular its cognitive-psychological
origin, and 2) the distinguished status of Euclidean geometry
relative to the totality of formally possible geometries. Both problems are closely related; the relation is roughly this. The manner
in which we recognize outer objects in space is by 'turning around'
and by 'moving towards' the horizon. Assuming 'free mobility of
rigid bodies' with respect to rotations and translations (as we
seem to experience them with our own bodies), this assumption
leads, according to the Helmholtz–Lie theorem, to the selection
of a particular class of spaces, namely, those of constant curvature. If we then assume that the linear translations form a normal
subgroup among the group of all continuous differentiable trans-

formations we arrive at the special Euclidean space, whose curvature equals zero.

The limits of this paper preclude the opportunity to explore many of the interesting details of Becker's proposal for a 'simultaneous' solution of both problems, the phenomenological constitution of space by specifying its cognitive-psychological origin, and the justification of the privileged status of Euclidean geometry by taking linear transformations as essential for our intuition of space as infinite. But at least three short remarks are appropriate to avoid possible misunderstandings. First, it is important to note that Becker, in complete agreement with Husserl, acknowledges the existence of a *multiplicity* of geometries in a formal mathematical sense. Second, and equally important, is that Becker claims a privileged status for Euclidean geometry only with respect to our 'natural intuition,' not with respect to physics.[9] The physicist is free to choose any geometry that he or she believes most appropriate for the description of the physical world. There are no methodological restrictions. Classical physics is, of course, based on Euclidean geometry. But this is only an *historical* fact, not a logical necessity, whose contingency has to be explained by an analysis of the epistemological reasons for this particular choice. Becker even agrees with Hermann Weyl that physics may choose, contrary to our natural intuition, a geometry with *variable* curvature, as Riemann had envisaged for the case that there is no free mobility of rigid bodies in space. (A possibility that had just become a matter of fact by the experimental vindication of General Relativity.) Last but not least, Euclidean geometry owes its singular *privileged* status, according to Becker, to the complete *homogeneity* of space, that is, that no point in space is qualitatively distinguished from any other point, a circumstance, which Becker thinks, is realized in our 'natural intuition.' The necessary condition for this is, mathematically speaking, connected with the circumstance that the translations form a normal *subgroup* of the congruent transformations (of rigid bodies). This consideration agrees with the fact, proved by Hilbert and Dehn, that the axioms of congruence and continuity are not sufficient for a foundation of Euclidean geometry: that there exist non-Euclidean geometries in which the axioms of congruence and continuity hold. Euclidean geometry not only requires the axioms of congruence and continuity but a further assumption that is equivalent to the axiom of parallels, i.e. that the sum of a triangle's internal angles equals the sum of two right angles.

From what I have said so far it seems obvious that the phenomenological point of view is much closer to the *group-theoretic* approach than to Hilbert's *axiomatic* point of view, at least in Becker's presentation of the matter. Yet a closer inspection reveals significant disagreements between the phenomenological and the group-theoretic approach of HLP. I will begin with the three most obvious points of agreement, before I consider the differences.

First, Becker makes absolutely no use of the axiomatic method. Instead, he refers the reader to Weyl's work in physics and differential geometry. It is well known that Weyl was for some time one of the two main critics of Hilbert's axiomatic approach.[10] Second, Becker uses mainly group-theoretic means in order to characterize the different types of perceptual spaces. In this regard he follows the tradition of HLP, although he does not begin with projective geometry but with Husserl's theory of recognition of objects in space. Third, and for our concern most important, Becker uses Poincaré's idea of *compensation* by motion to distinguish between two kinds of changes of form or the *gestalt* of an object: the changes caused merely by a shift of perspective and so can be compensated by a corresponding inverse motion of the perceiver, and the changes that cannot be compensated by motion and are in this sense *real* changes of the gestalt of an object. This distinction can then be used to distinguish rigid bodies from non-rigid objects. It is important to note that such a consideration only makes sense, if one takes the idea of rigid bodies and their motions as fundamental for a theory of space and, hence, for geometry. Otherwise the notion of compensation of gestalt changes by motion is of little significance and in no way fundamental for geometry.

The most interesting point is, however, that I was unable to find the concept of 'compensation' in Husserl's writings. This failure may be blamed on me.[11] I suspect that Becker had not only borrowed the notion of 'compensation' from Poincaré, but adopted the entire idea of *discriminating* rigid from non-rigid bodies through compensation of gestalt changes. In other words, Becker's presentation of the phenomenological position seems, in the decisive epistemological aspect, closer to Poincaré's view (that the idea of rigid bodies is fundamental for our knowledge of geometry) than to Husserl's original epistemological position regarding the *sources* of our knowledge of geometry. This initial suspicion is further supported by a more detailed examination of Husserl's writings on geometry.

Strictly speaking, there are only two geometrical papers, if we disregard for the moment the sections on geometry in *Krisis* and some minor pieces on the philosophy of space and intuition from his pre-critical period. The first paper is entitled "*Geschichtlicher Überblick über die Entwicklung der Geometrie*" and is part of the lecture on "Philosophy of Mathematics" in the winter term 1889–90. The second dates from 1892–93 and is entitled by the editor as "*Die Voraussetzungen [der Geometrie]*." The first paper entails what the title promises: an historical overview of the development of geometry from Euclid to Riemann. Towards the end it deals with Gauss's theory of surfaces. The second paper is mainly concerned with questions in the context of "*Streckenrechnung*" (i.e. the theory of line segments) and relies heavily on Pasch's "*Vorlesungen über neuere Geometrie*." Both papers would demand a detailed analysis, which is beyond the scope of the present essay. There is, however, a shorter paper, edited as "*Beilage VI*" on the history of geometry, which is very interesting and extremely relevant for our main question regarding Husserl's exact position between the group-theoretic approach of HLP and the axiomatic approach of Hilbert.

There is a certain difficulty with this text, which I shall explain before I draw any conclusions from it. The paper is dated by the editor ca.1900 and deals with two different "directions," as

Husserl calls them, in the foundation of modern geometry: first, the direction of Gauss, Lobatschewski, etc., coming from Euclid and second, the newer direction of Riemann, Helmholtz, and Lie, who understand the motions regarding the infinitesimal trans-formations of rigid bodies as the proper foundation of geometry. Husserl's paper reads like an excerpt from Hilbert's essay *"über die 'Grundlagen der Geometrie,'"* previously mentioned. This im-portant essay was published for the first time in 1902 and since reprinted as *Anhang IV* to the *"Festschrift"* in the *Grundlagen der Geometrie.*

This raises the difficult problem how to assess Husserl's text. Is it merely Husserl's *uncritical summary* of Hilbert's paper, or is it Husserl's *approval* of Hilbert's approach? Before drawing a hasty conclusion let us consider a third possibility that I contend comes closest to the real historical situation. Husserl's text is both a critical but nonetheless sympathetic appraisal of Hilbert's point of view as it is presented in the 1902 essay, and a definite rejection of the HLP approach. Let us consider some of Hilbert's remarks regarding the approaches to geometry. Hilbert opens his essay with the following extremely careful and polite but fundamen-tally critical remark.

> The investigations of Riemann and Helmholtz into the foundations of geometry compelled Lie to tackle the problem of the axiomatic treatment of geometry by *presupposing* the concept of a group. This led this sharp-witted mathematician to a system of axioms. He was able to prove, using his theory of transformation groups, that this system was sufficient for the construction of geometry.

Hilbert then continues:

> However, Lie always made the assumption in the justification of his theory of transformation groups that the transformation func-tions that define the group can be *differentiated*. For this reason it remains undiscussed in Lie's work as to whether the assumption of differentiability is unavoidable in the axiomatization of geometry

or whether the differentiability of the respective functions is purely a consequence of the group-concept (and the other geometrical axioms). Furthermore, according to his procedure, Lie is *forced* to state explicitly the axiom that the group of motions is generated by *infinitesimal* transformations. These assumptions can be expressed geometrically only in a rather *artificial* and *complicated* manner. Furthermore, they seem only to be enforced by Lie's *analytical method*, not by the problem itself.[12]

This is, of course, an extremely *critical* remark regarding Lie's group-theoretic approach concerning the foundations of geometry. More precisely, this is not so much an attack on the use of group-theory in geometry, but a critique of the intuition behind this approach. Because Hilbert shows in his paper that one can set up a system of axioms "which also rests on the concept of group, but which entails only simple and *geometrical perspicuous assumptions*, and in particular does in no way presuppose the differentiability of the functions, that mediate the motions."[13] In other words, the whole idea to characterize space by *infinitesimal transformations*, i.e. by rotations and translations of rigid bodies, presupposes too much from an axiomatic point of view and should be substituted by an approach that avoids the idea of motion and works with pure geometrical concepts and relations. This is exactly what Hilbert does in the paper. Without going into any of the technical details, two clarifying remarks are in order. First, one could object, as has Friedman,[14] that Hilbert himself still speaks of *motions* when he speaks of the 'group of motions' or the "functions that mediate the motions." That is correct, but this only a residue of the HLP approach, which Hilbert *should* have avoided in his new axiom system, because he does not use the concept of motion at all – only that of *transformations* , which is not the same. Indeed, Hilbert himself later corrected this mistake, when he – in his lecture on General Relativity – returned to his 1902 essay and substituted the concept of "motion" with that of "*Deckung*."[15] Second, the essay is a paradigm example of Hilbert's true genius. He criticizes the HLP approach by attacking it with its own weapons. Hilbert takes over the group-theoretic approach to geometry, apparently in the spirit of HLP, but at the same time he is able to avoid the assumption of differentiability of the transformation functions, which still seemed necessary in the eyes of Lie.

It is interesting that Husserl (commenting on Lie's assumption that "the motions form a continuous group, generated by infinitesimal transformations") is even more outspoken than Hilbert:

> This is complicated and *non-geometrical*. One has to know the entire theory with all its presuppositions, i.e. that motion is mediated by differentiable, indeed by analytical functions. Furthermore [the existence of] infinitesimal transformations is presupposed. An enormous number of things are presupposed in the other axioms ... These are all *non-geometrical* presuppositions, which are not articulated, which are irrelevant to geometry, and solely depend on the method. Lie simply insists on applying his method.[16]

Next, Husserl turns to Hilbert's alternative proposal to axiomatize geometry, as presented in the 1902 essay, which primarily uses Cantor's theory of point-sets and Jordan's theorem. The latter says that a continuous plane closed curve [without knots] divides the plane into an inner and an outer area. (No concept of 'motion' is presupposed in this approach, only that of 'continuity.') Husserl makes a number of critical remarks, raises some tricky questions, and offers several suggestions to improve Hilbert's proposal. My interpretation of this part of Husserl's text is that Husserl looks at Hilbert's proposal with sympathy, even though he makes a number of critical remarks.

When assessing these critical remarks one must bear in mind that Hilbert's proposal in the 1902 essay is by no means his own, axiomatic point of view, as presented in the *Festschrift*, but already a compromise between the group-theoretic approach of Lie and his own axiomatic stance. This is made crystal clear by the concluding statement of the essay, in which Hilbert stresses the crucial difference between the essay and the *Festschrift*. In the essay the axiom of continuity precedes all other axioms as it does in Lie's approach; in the *Festschrift* the order is exactly the opposite. Here the axioms of continuity – the Archimedean axiom and the axiom of domain completeness – are the last ones; first they turn the preceding axiom–system into a complete theory, as Hilbert calls it, i.e. a theory whose domain of objects cannot be expanded because of consistency. In modern terms, the axioms of continuity turn [Euclidean] geometry into a categorical theory, which has – up to isomorphism – only a single model.

In conclusion, for Husserl and his phenomenological approach to mathematics in general, and to geometry in particular, the metalogical concept of categoricity (which he calls the "absolute definiteness" of a of an axiom–system) was of greatest significance. In spite of this, it becomes more than probable that Husserl was much closer to Hilbert's axiomatic point of view, as presented in the *Festschrift*, than to the compromise proposed in the essay, not to mention the group-theoretic approach of HLP. If this conclusion is correct, it implies that Husserl's concept of 'spatial intuition' must be very similar, if not identical, to Hilbert's concept of 'spatial intuition.' To be sure, Husserl and Hilbert's concept of 'spatial intuition' is fundamentally different from the type of kinematic intuition employed by Helmholtz, Lie, and Poincaré in their group-theoretic foundations of geometry. However, to articulate the essence of Husserl and Hilbert's concept of 'spatial intuition' is another, more difficult task. I can only offer some hints and suggestions.

First, and most important, for Husserl and Hilbert, the space of our spatial intuition is the Euclidean space, as we assume it unconsciously in all our actions – including the recognition and identification of objects – in our everyday life (*Lebenswelt*).

Second, this spatial intuition (of Euclidean space) is not simply given, in a nativist sense of 'given,' but generated in a long process of *biological* and *cultural* evolution. The latter, cultural part is very important, because otherwise intuition would be a mere psychological notion, which it is definitively not, for Hilbert or for Husserl.

Third, despite being *generated* in a long and hitherto insufficiently understood process of biological and cultural evolution, our spatial intuition of things (in concordance with Euclidean geometry) is in a certain methodological sense *a priori*. We cannot change it *at will*, so to speak, but have to *rely on it* in all our mental and physical actions – including the construction of instruments – until we have a better 'adapted' geometry, that is a geometry in which the physical objects and events can be presented more simply and uniformly.

Fourth, although our spatial intuition is generated in conjunction with the perceptual spaces of our various senses (visual, tactile, acoustic), it is by no means identical with them (either with one of them or with their conjunction). Its epistemological status is of a more *transcendental* character in the sense that we presuppose that the space of our intuition, in which the outer objects appear to us, has a mathematically simple and definitive relational structure. This structure is for both Hilbert and Husserl the result of an 'idealization' that transcends in principle the domain of the empirically given.

Fifth, despite the transcendental character of our spatial intuition – and hence of Euclidean geometry – Husserl and Hilbert do not assume that Euclidean space is necessarily the space of physics. Nor do they assume that it can be saved *ad infinitum* from contradiction with experience by conventionalistic strategies, as Poincaré (and to a certain extent even Einstein) does. On the contrary, both presume that Euclidean geometry may entail certain '*anthropomorphic*' elements, such as an infinite spatial extension, which have no correlate in physics. It is the common task of the philosopher and the scientist to find out *which* these anthropomorphic elements are.

Notes

1   David Hilbert, *Grundlagen der Geometrie* (Leipzig: Teubner, 1899).
2   This has been collected and published as *Studien sur Arithmetik und Geometrie, Text aus dem Nachlass, 1866–1901*, hrsg. von I. Strohmeyer (den Haag: Martinus Nijhoff, 1988).

3 Furthermore, there is the chapter about the history of geometry in Husserl's book *Die Krisis der europäischen Wissenschaften*, which I'll ignore since it was written thirty years after Husserl and Hilbert stopped working on the foundations of geometry.

4 Oskar Becker, "Beitrage zur phänomenologischen Begründung der Geometrie un ihrer physikalischen Anwendungen," *Jahrbuch für Philosophie und phänomenologische Forshung* 6 (1923): 385–560.

5 "Die Geometrie bedarf – ebenso wie die Arithmetik – zu ihrem folgerichtigen Aufbau nur weniger und einfacher Grundsäze. Diese Grundsätze heißen Axiome der Geometrie. Die Aufstellung der Axiome der Geometrie und die Erforschung ihres Zusammenhanges ist eine Aufgabe, die seit Euklid in zahlreichen vortrefflichen Abhandlungen der mathematischen Literatur sich erörtert findet. Die bezeichnete Aufgabe läuft auf die logische Analyse unserer räumlichen Anschauung hinaus." Hilbert, *Grundlagen der Geometrie*, 1.

6 My further considerations are restricted to Hilbert's new axiomatic approach and exclude Pasch's peculiar empiricist approach, since he is (his pioneering work notwithstanding) only a transition figure. Something similar can be said of Klein. He articulated the programme but the decisive groundbreaking work was done first by Helmholtz and then later by Lie and Poincaré. In particular, Poincaré used the famous Helmholtz-Lie theorem according to which spaces of *constant curvature* cannot be discriminated just by the 'free mobility' of rigid bodies (or the notion of *congruence*, which holds in all spaces of constant curvature). On the basis of this theorem, Poincaré underpinned the group-theoretic approach with his conventionalism, which holds that the choice of a certain geometry is a matter of 'convention;' it is not based on empirical evidence. For this reason I will sometimes call the group-theoretic approach the HLP approach.

7 This is made clear by Hilbert towards the end of the 1902 paper "*Grundlagen der Geometrie*." Unfortunately, however, the paper of 1902 was not often read, at least not by the physicists of the time, because in general they followed the group-theoretic approach of HLP. The same point is reinforced in Hilbert's 1916–17 lectures about the foundations of physics, "*Die Grundlagen der Physik*." Here Hilbert, referring to the 1902 paper, crossed out "*Bewegung*" and "[*Bewegungs*] transformation," which he had still used in the 1902 paper, and replaced them with the concept of "*Deckung*" and "*Decktransformations*."

8 "Das erste Problem (die rationale Erfassung des Schlicht-Anschaulichen)

ist in der üblichen mathematischen Terminologie ausgedrückt, nichts anderes als das Kontinuumsproblem; das zweite (die Überwindung des
kontingenten Charakters) aber ist das Problem des Verhältnisses der eukli-
dischen zu den sog nicht-euklidischen Geometrien," Becker, *Beiträge zur
phänomenologischen Begründung der Geometrie*, 13.

9 This is my short term for Becker's lengthy and contorted expression '*der
Raum der schlicht anschaulichen Natur*,' that refers to the idea or representation of space that results from an unconscious cognitive process of our
visual and tactile systems.

10 Weyl knew, of course, very well what his former teacher had achieved in
geometry by the axiomatic approach.

11 It was suggested that in Husserl's book *Der Raum*, written about 1916 and
published posthumously, the notion of 'compensation' occurred with the
same meaning as Poincaré had introduced it. I was, however, unable to
verify this.

12 "Die Untersuchungen von Riemann und Helmholtz über die Grundlagen
der Geometrie veranlaßten Lie, das Problem der axiomatischen Behand-
lung der Geometrie unter *Voranstellung* des Gruppenbegriffs in Angriff zu
nehmen, und führten diesen scharfsinnigen Mathematiker zu einem System von Axiomen, von denen er mittels seiner Theorie der Transformations-
gruppen nachwies, daß sie zum Aufbau der Geometrie hinreichend sind.
Nun hat Lie bei der Begründung seiner Theorie der Transformations-
gruppen stets die Annahme gemacht, daß die die Gruppe definierenden
Funktionen differenziert werden können, und daher bleibt in den Lieschen
Entwicklungen unerörtert, ob die Annahme der *Differenzierbarkeit* [bei der
Frage nach den Axiomen der Geometrie] tatsächlich unvermeidlich ist
oder ob die Differenzierbarkeit der betreffenden Funktionen nicht viel-
mehr als reine Folge des Gruppenbegriffes [und der übrigen geometri-
schen Axiome] erscheint. Auch ist Lie *zufolge* seines Verfahrens genötigt,
ausdrücklich das Axiom aufzustellen, daß die Gruppe der Bewegungen
von *infinitesimalen* Transformationen erzeugt sei. Diese Forderungen ...
lassen sich geometrisch nur auf recht *gezwungene* und *komplizierte* Weise
zum Ausdruck bringen und scheinen überdies nur durch die von Lie
benutzte *analytische Methode*, nicht durch das Problem selbst bedingt."
Hilbert, *Grundlagen der Geometrie*, 178–79 (my emphasis).

13 Hilbert, *Grundlagen der Geometrie*, 185.

14 Michael Friedman, "Geometry, Construction and Intuition in Kant and his
Successors," *Erkenntnis* 42 (1995): 17.

15 See n.7.

16 "Das ist so kompliziert und *ungeometrisch.* [Man] Muß die ganze Theorie kennen mit all ihren Voraussetzungen. Z. B. daß die Bewegung vermittelt wird durch differenzierbare und sogar analytische Funktionen. Auch infinitesimale Transformation wird vorausgesetzt. Eine Unmenge von Dingen wird in den anderen Axiomen vorausgesetzt. ... Alles *ungeometrische* Voraussetzungen, die man nicht übersieht, Voraussetzungen, die nichts mit der Geometrie zu tun haben, sondern Voraussetzungen, die an der Methode haften. Lie will eben sein Methode anwenden" (my emphasis). Husserl, *Studien sur Arithmetik und Geometrie,* 412.

CHAPTER FIVE

Yvon Gauthier

HUSSERL AND THE THEORY OF MULTIPLICITIES "MANNIGFALTIKEITSLEHRE"

———

Husserl's idea of a general theory of deductive systems was motivated by the mathematical theory of multiplicities. Husserl mentions the names of Riemann and Grassmann as the originators of the concept of multiplicity and he adds the names of Lie, Hamilton, and Cantor.[1] However, it is Riemann's theory of varieties or n-dimensional spaces and Grassmann's extension theory that are the true sources of the concept. Husserl believes that he contributes to the theory with his notions of *plethoïd, orthoïd*, and cyclic multiplicities.[2] Though some real contribution from *Die Philosophie der Arithmetik,* where a general definition of multiplicity is attempted, might have been expected, we have rather a general calculus of operations that Husserl identifies with a "general arithmetic."[3] Although the expression is frequent in nineteenth century mathematics, it has a particular significance in the work of Kronecker and it is that aspect of the problem that will be addressed here.

1. GENERAL ARITHMETIC

Husserl mentions Kronecker mainly in his early text "*Versuche zur Philosophie des Kalküls*" in relation to general arithmetic.[4] It is

———

known that Husserl attended Kronecker's lectures in Berlin during the years 1878-1881, but he quotes only Kronecker's *"Ueber den Zahlbegriff"*[5] and never the major work of 1882, *"Grundzüge einer arithmetischen Theorie der algebraischen Grössen."*[6]

The manuscript *"Die Wahren Theorien"* (1889–90) contains Husserl's starting point in general arithmetic.[7] The *"arithmetica universalis,"* Husserl claims, is the same thing as *"mathesis universalis,"* which for Husserl, is pure logic or the theory of deductive systems, i.e. the theory of multiplicities. General arithmetic is the second source of Husserl's multiplicity theory and though Husserl thinks in terms of Hankel's permanence principle for formal laws, *"Prinzip der Permanenz der formalen Gesetze"* as he maintains in *"Formale und transzendentale Logik,"*[8] Kronecker speaks rather of conceptual determinations *"Begriffbestimmungen"* that are conserved in algebraic extensions. I offer here a sketch of Kronecker's general arithmetic.

Theory of forms, *'Formenlehre,'* is the name given by Grassmann and Hankel to the general theory of mathematical objects; however, the name has a more specific meaning in number theory from Lagrange and Legendre to Gauss and Kronecker. The theory deals with (homogeneous) polynomials and one can speak of quadratic forms like:

$$ax^2 + bxy + cy^2$$

which are homogeneous, i.e. the variables x and y have the same degree (exponent) and the constants a, b, and c are integer coefficients. It is the general theory of forms or polynomials that is the main topic of Kronecker's major work *Foundations of an Arithmetical Theory of Algebraic Quantities* and the introduction of a generalized notion of the indeterminate *'Unbestimmte'* is the principal ingredient of this general arithmetic of entire functions with integer coefficients.[9] The polynomial expression:

$$f(x) = a_0 x^n + a_1 x^{n-1} + ... + a_{n-1} x + a_n$$

with integer coefficients, a, and indeterminates, x, has the meaning of a finite support (due to the degree n) of an infinite formal power series – formal because convergence is not taken into account. Indeterminates have a more general function than functional variables, since they can be substituted for indefinite values (infinite or transcendent values as in Steinitz later on). Kronecker's constructivist goal consists in showing how far one can go in arithmetic with (algebraic) extensions. Kronecker's programme, which he realized for the most part, contains the field of rational functions (with integer coefficients) and its algebraic extensions. A general theory of *elimination* produces the polynomial decomposition of entire functions into irreducible factors and the algorithmic theory of divisors offers such a decomposition for the ring of integers and the field of algebraic forms in modular systems, i.e. of algebraic divisors: the linear representation (of degree 1) of algebraic quantities is given in terms of a finite number of elements. Hilbert admits that his celebrated theorem on a finite basis for a system of arbitrary forms is inspired by Kronecker's theorem, which states that in any field of algebraic functions, there is always a finite number of entire functions, such that any entire function can be represented by a linear function of that number. The Kroneckerian theory of indeterminates stands on the theory of equations (with the elimination of unknowns) initiated by Gauss and Galois; it introduces indeterminates only by association '*Associeren.*' The objective here is a proof of arithmetical existence of algebraic quantities, in Kronecker's words.[10] The process of association of indeterminates allows for the extension of the domain of arithmetic '*Gebietserweiterung der Arithmetik,*' which preserves the conceptual determinations of arithmetic, which we would call today the extensions of the 'natural' ring of polynomials including the field of algebraic numbers and the field of complex numbers in the algebraic closure of polynomial extensions.

The multiplicity theory Husserl briefly evokes in *Prolegomena* is a general theory of complex numbers encompassing the ordinary theory of complex numbers, real numbers, ordinal and cardinal arithmetic, vector analysis, and so on.[11] Although many contributed to the discussion including such names as Grassmann and Hankel, but also Cauchy, Weierstrass, Dedekind, and Cantor, there is no doubt that Husserl credits Kronecker with the achievement of what he calls "the arithmetic algorithm" in "*Die Wahren Theorien.*"

2. MULTIPLICITY THEORY OR "*MATHESIS UNIVERSALIS*" FROM *LOGISCHE UNTERSUCHUNGEN* TO *KRISIS*

Thirty-five years after *Logical Investigations*, Husserl still formulated his multiplicity theory in the terms of a '*mathesis universalis*' embodied in a formal logic of consistent multiplicities; among them are the definite multiplicities, '*definit Mannigfaltigkeiten*' the definition of which is given by a complete axiomatic formulation of a deductive system. Here Husserl refers to number theory or arithmetic and its algebraic extensions.[12] One can ask for the meaning of 'completeness' in that context. It is obviously neither Gödel's 1930 completeness theorem for the predicate calculus nor is it related to Gödel's 1931 incompleteness results. Instead, it is an axiom of completeness which is connected to Hilbert's completeness axiom for the arithmetic of real numbers in his consistency proof for Euclidean geometry in *Grundlagen der Geometrie*. It can be applied equally to Dedekind's theory of cuts for the real number system and it may be described as syntactic completeness rather than semantic completeness. With this in mind, it is evident that the model theory of algebraically closed fields or of real closed fields or of finite fields, that is axiomatizable and decidable theories which are complete, is a better analogy to Husserl's notion.[13] Another approach would link Husserl's theory of definite multiplicities with categoricity, but the fact that only second-order theories, like second-order Peano arithmetic, can be categorical, is not akin to Husserl's concept. The relevance of elementary or first-order algebraic theories to Husserl's theory is enhanced by the requirement that a definite multiplicity have the property of closure for which nothing remains indeterminate "es bleibt nichts mehr unbestimmt," as Husserl puts it in his *Ideen*.[14] Commentators, like J. Cavaillès, S. Bachelard, I. Strohmeyer, and L. Eley[15] seem to have missed this point completely and failed to consider Tarski's work, which builds upon Skolem and ultimately on Kronecker's theory of elimination, as shown by Van den Driess.[16]

3. THE FATE OF MULTIPLICITY THEORY IN HUSSERL'S THEORY OF SCIENCE

Husserl's programme of a "*Wissenschaftslehre*" develops into a transcendental phenomenology, and multiplicity theory occupies a central place in that development. From the early works to *Krisis*, the theme of the '*Mannigfaltigkeitslehre*' remains practically unchanged. '*Mathesis universalis*'[17] or pure logic comprising pure mathematics as a theoretical science,[18] multiplicity theory is seen by Husserl as a general theory which encompasses all deductive systems or theories. Husserl goes further in saying that multiplicity theory determines the form of all theories and their internal relationships. Such a theory of theories is conceived after the model of a '*Formenlehre*' theory of forms or polynomials of a general arithmetic, and if it has a mathematical motivation, it concludes nevertheless in a transcendental destination to the extent that it does not escape phenomenological reduction. Logic and mathematics are subject to bracketing '*Einklammerung*' or '*Ausschultung*,' as prescribed by the doctrine of *Ideen* since the pure forms of mathematics and logic cannot serve as tools in the material description of the intentional facts of transcendental consciousness.[19] There is a final limit to multiplicity theory: the εποχη of descriptive phenomenology with its unique ability to examine the phenomena of pure intuition. The same thesis is in *Formale und transzendentale Logik*: if multiplicity theory is the highest level of pure logic, it must nevertheless stop before the transcendental sphere as the idea of a formal ontology must be distinguished from a theory of science. Finally, transcendental logic must define the conditions of the possibility of a formal logic in the constitution of a transcendental egology, the last stage of Husserlian phenomenology which can only be grounded in the *'Lebenswelt'* or ante-predicative life.

Thus, the theory of science (deductive or nomological science) and its cornerstone, multiplicity theory, take a transcendental turn and leave open the question that I have posed elsewhere, "Is the theory of all possible theories at all possible?"[20] For Husserl there is no question of a TOE, *'theory of everything,'* or a unified field theory, but of a unitary theory of science. This ambitious programme has not been realized; it has been absorbed by transcendental phenomenology, which has had no significant impact on theoretical sciences, from logic to mathematics and physics. Multiplicity theory is a recurring theme in Husserl's work, remaining invariant and undergoing no major change from the beginning to the end of Husserl's intellectual life.

In sum, there is a major source for Husserl's project of a theory of theories: it is the general arithmetic of late nineteenth century mathematics whose main proponent was Kronecker. Although Husserl's project has not been brought to completion, Kronecker's programme of a general arithmetic has largely succeeded, but this is a different story.[21]

Notes

1 Edmund Husserl, *Formale und transzendentale Logik*, hrsg. von P. Janssen (den Haag: Martinus Nijhoff, 1974), 252.

2 Husserl, *Studien sur Arithmetik und Geometrie, Texte aus dem Nachlass* (1886–1901), hrsg. von I. Strohmeyer (den Haag: Martinus Nijhoff, 1983).

3 Husserl, *Philosophie der Arithmetik*. hrsg. von L. Eley (den Haag: Martinus Nijhoff, 1970), 493.

4 Husserl, *Studien sur Arithmetik und Geometrie.*

5 L. Kronecker, "Ueber den Zahlbegriff," in *Werke*, ed. K. Hensel, (New York: Chelsea, 1968), 2: 252–74.

6   Kronecker, "Grundzüge einer arithmetischen Theorie der algebraischen Grössen", in *Werke*, 3: 245–387.

7   Husserl, *Studien sur Arithmetik und Geometrie*.

8   Husserl, *Formale und transzendentale Logik*, 101.

9   Kronecker, "Grundzüge einer arithmetischen Theorie der algebraischen Grössen."

10 Kronecker, "Grundzüge einer arithmetischen Theorie der algebraischen Grössen," 296.

11 Husserl, *Logische Untersuchungen, Erste Band, Prolegomena zur reinen Logik*, hrsg. von E. Holenstein (den Haag: Martinus Nijhoff, 1978), 253.

12 Husserl, *Die Krisis des eurapäische Wissenschaften und die transzendentale Phänomenologie*, hrsg. von W. von Biemel, *Husserliana* Band VI, 2 *Auflage* (den Haag: Martinus Nijhoff, 1962), 44–45.

13 Husserl, *Formale und transzendentale Logik*, 31.

14 Husserl, *Ideen zu einer reinen Phänomenologie und phänomenologischen Philosophie*. Erste Buch, hrsg. von W. von Biemel (den Haag: Martinus Nijhoff, 1950), 167.

15 See J. Cavaillès, *Sur la logique et la théorie de la science* (Paris: PUF, 1947); S. Bachelard, *La logique de Husserl* (Paris: PUF, 1957); Husserl, *Studien sur Arithmetik und Geometrie*; Husserl, *Philosophie der Arithmetik*.

16 L. Van den Dries, "Alfred Tarski's Elimination Theory for Real Closed Fields," *Journal of Symbolic Logic* 53 (1988): 7–19.

17 Husserl, *Logische Untersuchungen*, 251.

18 Husserl, *Einleitung in die Logik und Erkenntnistheorie. Vorlesungen*, 1906–07, hrsg. von V. Melle (den Haag: Martinus Nijhoff, 1983).

19 Husserl, *Ideen*, 59.

20 Y. Gauthier, "La théorie de toutes les théories possibles est-elle possible?" Dialogue 14 (1) (1975): 81–87.

21 See Gauthier, *Logique interne* (Paris: Diderot, 2000).

CHAPTER SIX

Mathieu Marion

HUSSERL'S LEGACY IN THE PHILOSOPHY OF MATHEMATICS: FROM REALISM TO PREDICATIVISM

The list of mathematicians and philosophers of mathematics who claimed to have been influenced by Husserl is rather impressive. It includes Weyl, who successively claimed that his own predicativist programme in his 1918 book *The Continuum*, and a few years later, the intuitionism of Brouwer that he espoused at that stage were to be linked, as far as epistemology is concerned, to Husserlian phenomenology. In the preface to *The Continuum*, one reads:

> Concerning the epistemological side of logic, I agree with the conceptions which underlie Husserl's *Logical Investigations*. The reader should also consult the deepened presentation in Husserl's *Ideas Pertaining to a Pure Phenomenology and a Phenomenological Philosophy* which places the logical within the framework of a comprehensive philosophy.[1]

As is well known, Weyl abandoned his own programme in favour of Brouwer's intuitionism in the early 1920s. Although there are incompatibilities between his own predicativist programme and intuitionism, even in Weyl's own version, which differs on some not so trivial points from Brouwer's, he nevertheless saw some deep connections between intuitionism and phenomenology.[2] In

---

a paper in which he tried to adjudicate the *Grundlagenstreit* between Hilbert and Brouwer, Weyl wrote emphatically:

> If Hilbert's view prevails over intuitionism, as appears to be the case, then I see in this a decisive defeat of the philosophical attitude of pure phenomenology, which thus proves to be insufficient for the understanding of creative science even in the area of cognition that is most primal and most readily open to evidence – mathematics.[3]

Weyl was not the only one to see connections between intuitionism and phenomenology. So too, did one of Husserl's assistants, Becker, in his *Mathematische Existenz*,[4] which appeared in the same 1927 issue of Husserl's *Jahrbuch* as Heidegger's *Sein und Zeit*. Becker blended phenomenology and Heideggerian ideas. He influenced in turn Heyting, a student of Brouwer, who referred to Husserl's theory of intentionality in his well-known presentation of intuitionism to the Königsberg conference in 1930[5] and whose basic idea can be summarized as being that mathematical constructions are identified with fulfilled mathematical intentions.[6] In addition, the Viennese philosopher Kaufmann claimed the legacy of Husserl in his 1930 book, *The Infinite in Mathematics and Its Elimination*, where he also propounded a form of mathematical constructivism – more precisely a form of finitism which is not equivalent to intuitionism, since he did not call for a rejection of the universal applicability of the Law of Excluded Middle, because he believed, erroneously, arithmetic to be decidable.[7] Gödel's incompleteness results invalidated his standpoint a year later.

On another note, Cavaillès linked Husserl's phenomenology with the formalist programme of Hilbert, Brouwer's arch-rival, making heavy weather of the connection between Husserl's notion of 'definiteness' with Hilbert's axiom of completeness; a connection made by Husserl himself in *Formal and Transcendental Logic*,[8] in connection with his *Mannigfaltigkeitslehre*.[9] Cavaillès thought that, although the notions of completeness involved here are different ones, Gödel's incompleteness theorems dealt a fatal blow to Husserl's phenomenology.[10] His idea that a

Husserlian 'philosophy of consciousness' had to be replaced by a (structuralist) philosophy or 'dialectic' of concepts had a lasting influence on French philosophy of mathematics – one thinks, for example, of Desanti, who developed this structuralist approach in *Les idéalités mathématiques*.[11] Linking phenomenology with Hilbert's programme contradicts Weyl's opinion in the passage just quoted. I shall not enter this debate; rather, I shall state without argument that both claims are wrong: Husserl's phenomenology should be linked neither with either Hilbert's programme, nor with Brouwerian intuitionism, nor, for that matter, with any constructivist programme. As a constructivist, I am far from being convinced that phenomenology can serve as the basis for a clear and convincing argument in favour of the abandonment of classical mathematics.

In light of these constructivist and structuralist claims to Husserl's legacy, the fact that Gödel, who is usually considered to have propounded the strongest form of realism about mathematical entities in the twentieth century, also claimed connections with phenomenology will appear at first blush as a rather wild one. In his Gibbs Lecture in 1951, Gödel described his own realist position in the following terms:

> I am under the impression that after sufficient clarification of the concepts in question it will be possible to conduct these discussion with mathematical rigour and that the result then will be that ... the Platonistic view is the only one tenable. Thereby I mean the view that mathematics describes a non-sensual reality, which exists independently both of the acts and of the dispositions of the human mind and is only perceived, and probably perceived very incompletely, by the human mind.[12]

It appears that Gödel only started reading Husserl's major works some years later, in 1959.[13] Gödel was quite taken by Husserl's phenomenology, which he described, in an unpublished paper dating from 1961, as avoiding "both the death-defying leaps of idealism into a new metaphysics as well as the positivistic rejection of all metaphysics."[14] To increase the confusion over Husserl's legacy, Gödel also reinterpreted, with hindsight, his own incompleteness theorems in light of phenomenology, thus contradicting the claim made by Cavaillès that these theorems dealt a decisive blow to phenomenology. According to Gödel, his own results meant that

> the certainty of mathematics is to be secured not by proving certain properties by a projection onto material systems – namely, the manipulation of physical symbols – but rather by cultivating (deepening) knowledge of the abstract concepts themselves which lead to the setting up of these mechanical systems, and further by seeking, according to the same procedures, to gain insights into the solvability, and the actual methods for the solution, of all meaningful mathematical problems.[15]

Gödel further believed that one could not get such insight into these abstract concepts by trying to give explicit definitions since "one obviously needs other undefinable abstract concepts and axioms holding for them."[16] Therefore, Gödel concludes, "the procedure must consist, at least to a large extent, in a clarification of meaning that does not consist in giving definitions."[17] And then Gödel claims,

> there exists today the beginning of a science which claims to possess a systematic method for such a clarification of meaning, and that is the phenomenology founded by Husserl. Here clarification of meaning consists in focusing more sharply on the concepts concerned by directing our attention in a certain way, namely, onto our own acts in the use of these concepts, onto our powers in carrying out our acts, etc.[18]

In his influential 1947 paper, "What is Cantor's Continuum Problem?", Gödel pleaded for the discovery of new axioms for so-called large cardinals, in order to decide the continuum hypothesis.[19] Much work in set theory in the 1960s and 1970s in particular was influenced by Gödel's paper.[20] In my opinion it turned out, to put it bluntly, to be more of a wild goose chase than anything else. But it turns out that Gödel actually thought, after studying intensely Husserl's writings in 1959–1960, that one could profitably use the phenomenological method in order to frame those new axioms.

My brief survey has indicated some wild discrepancies in the claims to Husserl's legacy: his name has been used to countenance both revisionist, constructivist programmes, on the one hand, and non-revisionist programmes aimed at preserving and extending classical mathematics on the other hand. So either Husserl's phenomenology is fundamentally ambiguous and it is of no help in trying to adjudicate any debate about the foundations of mathematics, since it appears to justify at the same time about all possible positions under the sun, or some of the above-mentioned authors have illegitimately appropriated Husserl's phenomenology. Since charity forbids one to begin with the first alternative, that of ruling out phenomenology as ambiguous, one should rather ask: Who could really lay claim to the legacy of Husserl?

It is impossible within the limits of this paper to present my answer to this question *and* present in support of it all the necessary arguments and counter-arguments, let alone a full discussion of the literature involved. Consequently, I shall focus on two authors: Weyl, in his predicativist phase, and Gödel. There can be no surprise in my answer: Weyl's predicativism is indeed generally based on 'phenomenological' considerations but it contains essential departures from Husserlian orthodoxy in these matters – and the motives for these are to be found elsewhere – while Gödel's realist stance has truer affinities with it. In other words, Husserl's phenomenology should be seen as countenancing a form of realism not unlike that propounded by Gödel.

Again, due to space restrictions, I shall leave aside the issue of possible connections between phenomenology and intuitionism. I shall merely justify it by stating that it appears to me that in Weyl's case, the connections that he drew between intuitionism and phenomenology are less detailed than those which he drew with his own predicativist programme. Much of the impetus for the connections between intuitionism and phenomenology comes from the work of Becker, which, alas, I must leave aside.[21] It is clear, at least, that there are few historical connections between Husserl and Brouwer.[22] At any rate, there seem to me to be *prima facie* unbridgeable differences between their respective standpoints, for example concerning time, since for Husserl mathematical entities are timeless and they certainly are not for Brouwer.[23]

Before presenting my case, I should like to say a few words about a potential ambiguity in Husserl's phenomenology. I am not going to claim that Husserl's phenomenology is entirely free from ambiguities but there is at least one such ambiguity that I should like to claim as irrelevant to my argument. Ever since Ingarden's detailed criticisms, we are all aware of the implications for the destiny of phenomenology of Husserl's so-called turn towards "transcendental idealism,"[24] which is said to have occurred roughly between *Logical Investigations* and *Ideas* I. There have been substantial disagreements on this issue in the secondary literature; I should simply state that I side with those who see the transcendental turn as a wrong turn.[25] In any case, I have in mind one possible way out of our dilemma about Husserl's true legacy, i.e., the idea that the transcendental turn might explain its ambiguity: someone might simply think that Weyl, the constructivist, picked on the more subjectivist, Kantian aspects of Husserl's phenomenology after the 'turn,' while Gödel was more faithful to the earlier realism of *Logical Investigations*.

This convenient solution, according to which both authors have an equal claim to the legacy of Husserl, simply won't do. To begin with, it is not clear how much of the consequences of this turn were properly understood by Weyl or Gödel, whose readings of Husserl, incidentally, certainly feature Kantian elements that are not really present in *Logical Investigations*. It is also noticeable that the primary influence on *The Continuum* is indeed *Logical Investigations*, while Gödel was, according to Føllesdal, "more appreciative of the *Ideas* I and of other works written after

Husserl's 'idealist' conversion around 1907."[26] I should like to claim further that the transcendental turn simply leaves the issue at hand undisturbed.

By this I mean the following. When Weyl sent a copy of *The Continuum* to Husserl, the latter wrote back describing the book as a "significant event" for himself and expressing his delight at the fact that, at last, a mathematician showed some "understanding of the necessity of phenomenological modes of investigation for the clarification of fundamental notions, and who thus finds himself on the fundamental ground of the logico-mathematical intuition."[27] I wish to focus in what follows on this very notion of "logico-mathematical intuition", which is taken straight out of *Logical Investigations*. My claim is simply that the transcendental turn left the important elements of this notion essentially untouched.

To see this, I need to explain Husserl's basic position on the "logico-mathematical intuition."[28] In chapter one of the First Logical Investigation, Husserl distinguished between the meaning of an expression and the 'objective correlate' or 'objectivity' (*Gegenständlichkeit*, *Objektivität*) referred to by the expression. In this same chapter, section nine, Husserl distinguished between meaning-conferring and meaning-fulfilling acts. But it is only in the last, sixth investigation that Husserl discussed in detail issues related to the latter. With respect to the formal constituents of statements such as the particles 'is,' 'and,' 'or,' 'not,' 'all,' 'some,' etc. and with respect to numerical terms and relational expressions such 'greater than,' 'next to,' etc., Husserl argued in chapter six of the Sixth Investigation that, although they do not obtain their fulfilment in sensuous perception and although intuitive counterparts of these cannot be represented sensibly in the imagination, it does not mean that the meanings of these formal constituents cannot be fulfilled by some corresponding objectivities. Accordingly, Husserl argued in § 45 that there must be, alongside sensuous intuition some *analogous* act of 'supersensuous' intuition, which he calls "categorial" intuition and by means of which one intuit the objectivities that fulfil these meanings. (The analogy between 'sensuous' and 'categorial' intuition will become important below, when discussing Gödel on mathematical intuition.) Acts of categorial intuition, on which objects such as collections or sets, numbers, and states of affairs are grasped, are said to be based or 'founded' on prior acts of sensuous intuition. Categorial acts may also serve as founding acts of new categorial acts of higher level, for example, when one establishes a relation between two sets. Since categorial acts of, say, second level can serve as founding acts of categorial acts of third level, and so on, a hierarchy of types of categorial acts obtains, with acts of sensuous intuition a zero level and categorial acts of level one whose founding acts are all acts of sensuous intuition of level zero, and so on. Although acts of categorial intuition are based on acts of sensuous intuition, they are not reducible to them: in an act of sensuous intuition the objects are simply apprehended, but in a categorial act the objects are grasped in founded acts that connect what is given in the founding sensuous act. They are objectivities of a higher level and of a different sort: non-sensuous objectivities or, in our contemporary jargon, abstract objects. Objectivities that are grasped in categorial, founded acts can, however, only be given in the founding acts but, as pointed out in § 61 of the Sixth Investigation, they do not amount to a transformation of sensibly given objects or to new such objects, not already contained as parts in the founding acts.

At the end of § 45 of the Sixth Investigation, Husserl distinguishes categorial intuition "in the narrower sense," where "perception terminates upon an individual," from the "widest sense" where "universal states of affairs can be said to be perceived ('seen', 'beheld with evidence')." According to § 52 of the Sixth Investigation, the latter refers to a further type of categorial acts, "Ideational Abstraction," where the "Idea," the "Universal" is "brought to consciousness, and achieves *actual givenness.*" Husserl wrote univocally about an "intuition" or "perception" of universals – there is no room in that text for a Kantian form of "constitution"; one may properly speak of an "intuition of essences." This is the source of the notion of *Wesensschau.*[29]

What about logical and mathematical entities? Husserl further distinguishes in § 60 of the Sixth Investigation between sensuous abstraction and pure categorial abstraction. Sensuous abstraction gives us purely sensuous concepts such as 'house,' 'red,' etc., and mixed concepts, where sensuous concepts are mixed with formal ones, such as the concept of colouring or the parallel axiom; while pure categorial abstraction gives us purely categorial concepts such as the concept of 'relation,' 'set,' and 'number.' The class of such concepts is called by Husserl "formal-ontological categories." Purely sensuous and mixed concepts have their ultimate foundation in sensuous intuition, while categorial concepts have their foundation in categorial intuitions. To give only one example, given a categorial intuition of a set, pure categorial abstraction directs itself to the form of the collection, leaving aside everything material in the members of the set, i.e., leaving them completely indeterminate. Thus, the 'logico-mathematical intuition' is categorial intuition purified, so to speak, by pure categorial abstraction and pure logic and mathematics do not contain any sensuous concept in their foundation. Once the primitive mathematical concepts, i.e., the formal-ontological categories, are thus grasped, further entities may be grasped in new categorial acts of higher level and the whole mathematics is thus built on the basis of the formal-ontological categories that are grasped in pure categorial abstraction, which leaves indeterminate the sensuous material on which a categorial intuition of first level was based.

Little of the foregoing has changed by the time Husserl returns to the issue in one of his last projects, the posthumously published *Experience and Judgement*, e.g., §§ 60–61.[30] In this book, he added, however, that in categorial acts the object is never apprehended in a purely receptive manner as it is in acts of sensuous intuition and that it requires the intervention of the spontaneity of understanding; this is why he called the objects given in categorial intuition "objectivities of understanding."[31] Although Husserl thus spoke of "objectivities of understanding," his analysis of mathematical objectivities, which is the only aspect important to my argument, *remained in essence the same*. It is true that, in a letter to Ingarden, Husserl made it plain that, as far as he was concerned, his own transcendental turn had to do with the abandonment of the (Platonistic) notion of ideal meaning as "species," which was introduced in § 31 of the first of the *Logical Investigations*. But Husserl was keen, in § 64 of *Experience and Judgement*, on distinguishing between "objectivities of understanding" and "species," obtained by generalization.[32]

It should be clear by now that the transcendental turn cannot be made to account for the diverging claims of Weyl and Gödel, as it simply makes no practical difference as far as the 'logico-mathematical intuition' is concerned. One must now look at their conceptions and see if they are faithful or not to this 'logico-mathematical intuition.'

Husserl's influence on Weyl is clearly visible in passages such as section six of chapter two of *The Continuum*, where Weyl provides a description of the intuitive notion of the continuum that is based on the description of (the intentional constitution of) phenomenal time in §§ 81–2 of *Ideas* I. One also finds in that book statements that show that Weyl had adopted a clearly idealistic reading of Husserl, in line with the transcendental idealism of *Ideas* I. For example, Weyl claims in the same chapter that "existence is only given and *can* only be given as the intentional content of the conscious experiences of a pure, sense-bestowing I."[33] This idealism has been credited by some commentators to the influence of Fichte, whom Weyl was fond of quoting and whose influence is far from being superficial; it comes clearly to the fore in later texts such as "Insight and Reflection."[34] I shall not discuss this point further. Such appropriations of phenomenology should not attract attention away from the fact that his predicativist programme in *The Continuum* diverges in a number of fundamental ways from Husserl. I shall merely focus here on two major structural changes, which, to my mind, seriously undermine the claim that Weyl's predicativism may represent a 'logico-mathematical' continuation of Husserl's ideas.[35]

The first change concerns level zero of the hierarchy, where, according to Weyl in *The Continuum*, the whole sequence of the natural numbers is given in an immediate, non-founded "intuition of iteration, i.e., of the infinite repetition of a procedure."[36] As is well-known, according to Husserl there cannot be direct access to formal-ontological categories, presumably, the sequence of the natural numbers would be one of these; one can only get to them in founded categorial acts.[37] This significant departure is probably due to the influence on Weyl of Poincaré, who thought that the natural numbers were given to us by some *a priori* synthesis.[38] One important aspect of this influence is the fact that Weyl thus assumes that one can intuit an actually infinite series of steps. This is a clear departure from Husserl, for whom infinite and even finite sets can be at best only inadequately or partially intuited.[39] On this very point, I shall claim that Gödel was more faithful to Husserl.

The second change concerns all higher level objectivities whose constitution presupposes the basic domain of natural numbers, as given in immediate intuition. Although the latter are thus conceived as objectivities that exist independently, every other objectivity to be constituted at any higher level clearly is not, according to Weyl, since they are subjected to logical strictures. These logical restrictions are spelled out in section six of chapter one of *The Continuum*. Without going into the details of Weyl's programme, let me simply note that Weyl's restriction to predicative constructions which have as a basis the fundamental domain of the natural numbers was devised in order to avoid the well-known Russellian paradox and that here, once more, one feels the influence of Poincaré, who also led Russell to coin his "vicious-circle principle."[40] There has never been room for such logical restrictions, nor for any justification of them, within Husserl's phenomenology. He believed that mathematics was a form of knowledge of real as well as a purely formal theory of possible realms of being. As a matter of fact, I simply know of no passage where Husserl straightforwardly supported a restriction of mathematical knowledge in order to fit it into strictly intuitive (or predicative, for that matter) boundaries; on the contrary, he was much interested in the study of formal, non-interpreted systems – I am referring here to his *Mannigfaltigkeitslehre* – a study which formed part of his "formal ontology." This is the point behind the doctrine of the *Wesensschau*, this "great performance which is called the seeing of an *a priori*," as Husserl once wrote.[41] Phenomenology does not rule out purely formal, non-intuitive empty forms of mathematics, which would be deemed non-constructive, and Husserl certainly never opposed the development of formal or axiomatic theories of analysis as description of *possible* realms of being, with the *Wesensschau* grounding, so to speak, the 'theory of all theories,' i.e., the *Mannigfaltig-keitslehre*.[42] On this point, again, Gödel's use of phenomenology appears to have been more in line with Husserl's ideas.

One important aspect of these restrictions is that they imply that Weyl rejects Husserl's notion of *epoché*, since he clearly does not suspend ontological commitments. As Da Silva pointed out,[43] this rejection of the *epoché* is also evident in Weyl's book on *Space-Time-Matter*, published in the same year (1918) as *The Continuum*, in which Weyl doubts that there could be an intuition of space unaided by the constructions of mathematical physics. This is a straightforward denial of the possibility of an intuition of essences, within the scope of *epoché*.

This second departure from Husserl leads Weyl to the claim, in section six of chapter two of *The Continuum*, that there cannot be a phenomenological foundation of the (classical) mathematical continuum on the basis of an immediate intuition of phenomenal time analogous to that which gives us access to the natural numbers:

> The category of natural numbers can supply the foundation of a mathematical discipline. But perhaps the continuum cannot, since ... as basic a notion as that of the point in the continuum lacks the required support of intuition. It is to the credit of Bergson's philosophy to have pointed out forcefully this deep division between the

world of mathematical concepts and the immediately experienced continuity of phenomenal time.[44]

There is thus an unbridgeable gap: "the intuitive and the mathematical continuum do not coincide; a deep chasm is fixed between them" and Weyl proposes a "construction" of the continuum "which must establish its own reasonableness."[45] In other words, what Weyl is telling us is that a phenomenological theory of the continuum is not truly possible and that, in the end, his mathematical construction goes a few steps further than a mere phenomenological description.[46] One should also notice that, on this essential point of departure from Husserl's phenomenology, Weyl acknowledges the influence of Bergson.

I should like to insist on two of the above points: Weyl's claim about the intuition of iteration and of the sequence of natural numbers and his claim about the "deep chasm" between the intuitive and the mathematical continuum are points where he clearly demarcates himself from Husserl on the issue of the 'logico-mathematical intuition.' I could therefore conclude my remarks on *The Continuum* by agreeing with Da Silva, when he stated that "despite its debt to certain phenomenological ideas, the system of *The Continuum* cannot be seen as a prototype of how the whole of mathematics should be developed from the phenomenological perspective"[47] and that "despite inviting an analysis in terms of phenomenological notions" this book "cannot be seen as a natural mathematical development of ... Husserlian ontology and epistemology."[48]

As has been discussed above, Gödel seriously studied Husserl's major works after 1959, therefore many years after the last of his major logical results. One cannot, therefore, credit phenomenology with the discovery of his famous results; his remarks about phenomenology would thus appear irrelevant, since they evidently played no role in the finding of his results. This is, however, incorrect in a limited but important sense, since *Gödel's interest in Husserl was actually sparked by the strong affinities that he saw with his own realism* as stated in the earlier quotation from the Gibbs lecture. In a letter to Hao Wang, Gödel argued that it was because he held this very realist or "objectivist conception of mathematics," as he called it, that he looked for and obtained his famous completeness and incompleteness results, while others, such as Skolem, working within the finitist constraints of Hilbert's programme, could not have looked for such results – he even claimed that his 'objectivist' conception was at the basis of his other results.[49] When Gödel extensively studied Husserl after 1959, he was trying to find a refined, sophisticated version of this 'objectivist' conception and, in the process, he brought it in line, in a strikingly natural manner, with Husserl's phenomenology. There is no space to discuss all aspects of the influence of Husserl on Gödel in his attempt at framing his philosophical position. My claim will be limited to one central feature, that of mathematical intuition.

One finds statements of Gödel's 'objectivist' outlook in the 1950s, therefore prior to his reading Husserl. For instance, he attacked the conventionalist approach to the foundations of mathematics, writing:

> What is wrong [with conventionalism] is that the meaning of the terms (that is, the concepts they denote) is asserted to be something manmade and consisting merely in semantical conventions. The truth, I believe, is that these concepts form an objective reality of their own, which we cannot create or change, but only perceive and describe.[50]

After studying Husserl, Gödel presented a refined version of his take on mathematical intuition. In a supplement to "What is Cantor's Continuum Problem?" written in 1963, he wrote the now famous:

> despite their remoteness from sense experience, we do have something like a perception of the objects of set theory, as is seen from the fact that the axioms force themselves upon us as being true. I don't see any reason why we should have less confidence in this

kind of perception, i.e. in mathematical intuition, than in sense perception, which induces us to build up physical theories and to expect that future sense perceptions will agree with them and, moreover, to believe that a question not decidable now has meaning and may be decided in the future. The set-theoretical paradoxes are hardly any more troublesome for mathematics than deceptions of the senses are for physics. That new mathematical intuitions leading to a decision of such problems as Cantor's continuum hypothesis are perfectly possible was pointed out earlier.[51]

He added that "the question of the objective existence of the objects of mathematical intuition ... is an exact replica of the question of the objective existence of the outer world."[52] One should notice here the strong emphasis on the fact that mathematical intuition is analogous to physical perception. In another unpublished essay in which he critically discussed Carnap's conventionalism, written before 1959, Gödel was already clear on this point:

> The similarity between mathematical intuition and a physical sense is very striking. It is arbitrary to consider 'this is red' as an immediate datum, but not so to consider the proposition expressing modus ponens or complete induction .... For the difference, as far as is relevant here, consists solely in the fact that in the first case a relationship between a concept and a particular object is perceived, while in the second case it is a relationship between concepts.[53]

Again, this is similar to Husserl's view and when one keeps in mind the parallels with Husserl's own notion of categorial intuition as analogous to and as a 'widening' of sensuous intuition, *such passages do not appear anymore as the expression of a 'naïve' realism.*

There are other noteworthy similarities between Gödel's 'perception' or 'mathematical intuition' and Husserl's 'logico-mathematical intuition.' To begin with there is the idea that 'axioms force themselves upon us as being true.' According to Husserl, one cannot will objectivities of higher level at all. This is the point of his notion of categorial intuition, which is often obscured by idealistic readings of the 'constitution.' Furthermore, Gödel believed that mathematical intuition is not infallible, in face of the "inexhaustibility of mathematics;" this was the point behind his reinterpretation of the significance of his incompleteness theorems.[54] Husserl also believed that categorial intuition is not an infallible source of evidence. It always involves anticipations concerning aspects of the objects that have not been explored and which, in turn, might turn out to be wrong. This further parallelism accounts for Gödel's belief that one could use the phenomenological method precisely to find out new axioms that would help prove or disprove the continuum hypothesis. Finally, one should also recall here Gödel's claim that phenomenology is a "systematic method for such a clarification of meaning."[55] Clarification of meaning was seen by both Husserl and Gödel as the method to find out about mathematical objects. These striking parallels show, therefore, that Gödel had a very good grasp of Husserl's philosophy. As Føllesdal wrote, perhaps a bit over-enthusiastically: "Gödel's understanding of Husserl is extraordinary, placing him among the foremost interpreters of Husserl."[56]

I should like to conclude my remarks about Gödel by pointing out that his philosophy is often seen as a naive, unsophisticated form of realism. To my mind, as for other commentators such as Føllesdal or Tieszen, the strong connections between Husserl's phenomenology and Gödel's realism indicate that the latter is more sophisticated than usually thought. There is a strong Kantian flavour to Husserl's ideas, especially after the transcendental turn; his later philosophy was a genuine, if not successful, attempt at overcoming the usual realism–idealism debate. On the one hand, as I observed, the objectivities that are intuited are 'transcendent' in the sense that they are inexhaustible. Moreover, these objectivities are always experienced as independent of us. This is the Platonistic or realist aspect in Husserl. On the other hand, Husserl insisted on the fact that we do not play a merely passive role and that we are somewhat active in structuring reality so that, although these objectivities are seen as independent of us, we should not forget our role in the very process of 'constitution.' This is the Kantian part. Regarding the interplay of Kantian and Platonistic elements in Husserl's philosophy, I will only state that it is present in Gödel's as well. This is why attempts should be made to understand his philosophy in a more sophisticated way than before.[57]

To return to the subject of ambiguity in Husserl, I ruled out the potential role of the transcendental turn in explaining the diverging claims to his legacy. Perhaps this interplay of Kantian and Platonistic themes does constitute a fundamental ambiguity in Husserl's philosophy. It is clear for philosophers who are not inclined towards Platonism will insist on the Kantian themes and overlook or deny the Platonist ones. Perhaps attempts at legitimizing constructivist programmes from Husserl's philosophy derive entirely from focusing on these Kantian themes. I simply think that this would be a one-sided reading and that overlooking the Platonistic themes is to disfigure phenomenology. Attempts to link constructivist programmes with Husserlian phenomenology founder on precisely this point.

I should like to end with a few caveats. To begin with, I must insist on the fact that *my claim was not that Gödel is on the whole more faithful to Husserl but the more limited claim that in his discussion of mathematical intuition Gödel comes closer indeed to Husserl than Weyl ever did.* Moreover, having made this claim, I must insist on the fact that Gödel's discussion of the 'logico-mathematical intuition' appears to be much less detailed than Husserl's, that it does not add any new elements to it, and that it does not help us, for that matter, solve any of the problems facing Husserl. In short, Gödel may have understood the gist of Husserl's *Wesensschau;* the few *written* remarks that he has left us are a bit too shallow to be of any use. I do not defend Husserl's eidetic intuition – on the contrary, I point out that Gödel made more faithful use of Husserl's ideas, and the fact that not much came out of it could be seen as a *reductio ad absurdum* for Husserl's notion; to quote him on the very idea of science of essences, "the dream is over."[58]

Notes

1   Hermann Weyl, *The Continuum: A Critical Examination of the Foundations of Analysis* (New York: Dover, 1987), 2.

2   On the differences between Weyl's intuitionism and Brouwer's, see Ulrich Majer, "Zu einer bemerkenswerten Differenz zwischen Brouwer und Weyl," in *Exakte Wissenschaften und ihre philosophische Grundlegung: Votrāge des Internationalen Hermann-Weyl-Kongresses, Kiel 1985*, ed. W. Deppert, K. Hübner, A. Oberschelp, and V. Weidemann, 543–552 (Frankfurt: Peter Lang, 1988); Mark van Atten, *Phenomenology of Choice Sequences*, Quaestiones Infinitae 31 (Utrecht: Zeno, The Leiden-Utrecht Institute of Philosophy, 1999); Dirk van Dalen, "Hermann Weyl's Intuitionistic Mathematics," *The Bulletin of Symbolic Logic* 1 (1995): 145–69.

3   Weyl, "Comments on Hilbert's Second Lecture on the Foundations of Mathematics," in *From Frege to Gödel: A Sourcebook in Mathematical Logic, 1879–1931*, ed. Jean van Heijenoort, 480–484 (Cambridge MA: Harvard University Press, 1967).

4   Oskar Becker, "*Mathematische Existenz: Untersuchungen zur Logik und Ontologie mathematischer Phänomene*," *Jahrbuch für Philosophie une phänomenologische Forschung* 8 (1927): 439–809.

5   Becker, "*Mathematische Existenz*" section four; Arend Heyting, "The Intuitionist Foundations of Mathematics," in *Philosophy of Mathematics: Selected Readings*, ed. Paul Benacerraf and Hilary Putnam, 2d ed., 52–61 (Cambridge: Cambridge University Press, 1983).

6   This idea is further discussed in Richard Tieszen's "Mathematical Intuition and Husserl's Phenomenology," *Noûs* 18 (1984): 395–421 and "Phenomenology and Mathematical Knowledge," *Synthese* 75 (1988): 373–403.

7   Felix Kaufmann, "The Infinite in Mathematics and its Elimination," in *The Infinite in Mathematics: Logico-Mathematical Writings*, ed. Brian F. McGuinness, 1–164 (Dordrecht: D. Reidel, 1978), 153.

8   Husserl, *Formal and Transcendental Logic* (The Hague: Martinus Nijhoff, 1969), § 31.

9   For more on 'definite manifolds,' see Husserl, *Logical Investigations* (London: Routledge & Kegan Paul, 1970) (*Prolegomena*), § 69, 239ff., Husserl,

Ideas Pertaining to a Pure Phenomenology and to a Phenomenological Philosophy. First Book (The Hague: Martinus Nijhoff, 1983), § 72. See also Da Silva's "The Many Senses of Completeness," (forthcoming). For a clear statement of Cavaillès's opinion that Husserl's philosophy has nothing to do with Brouwer's intuitionism but is rather to be linked with Hilbert's programme, see Jean Cavaillès, *Méthode axiomatique et formalisme* (Paris: Hermann, 1981), 33n. See also Jules Vuillemin, *La philosophie de l'algèbre* (Paris: Presses Universitaires de France, 1962), 476–518, in particular 495.

10 Cavaillès, "On Logic and the Theory of Science," in *Phenomenology and the Natural Sciences: Essays and Translations*, ed. Joseph J. Kockelmans and Theodore J. Kisiel, 353–409 (Evanston, IL: Northwestern University Press, 1970), 405–9.

11 Jean-Toussaint Desanti, *Les idéalités mathématiques* (Paris: Éditions du Seuil, 1968). For a different standpoint, see Yvon Gauthier, *Fondements des mathématiques: Introduction à une philosophie constructiviste* (Montréal: Presses de l'Université de Montréal, 1976), chapter 9.

12 Kurt Gödel, *Collected Works III: Unpublished Essays and Lectures*, ed. S. Feferman et al. (Oxford: Oxford University Press, 1995), 322–3.

13 Hao Wang, *Reflections on Kurt Gödel* (Cambridge MA: MIT Press, 1987), 28, 46, 74, and 121.

14 Gödel, *Collected Works III: Unpublished Essays and Lectures*, 387.

15 Gödel, *Collected Works III: Unpublished Essays and Lectures*, 383.

16 Gödel, *Collected Works III: Unpublished Essays and Lectures*, 383.

17 Gödel, *Collected Works III: Unpublished Essays and Lectures*, 383.

18 Gödel, *Collected Works III: Unpublished Essays and Lectures*, 383.

19 Gödel, *Collected Works II: Publications 1938–1974*, ed. S. Feferman et al. (Oxford: Oxford University Press, 1990), 181–3.

20 See Penelope Maddy, "Believing the Axioms," *Journal of Symbolic Logic* 53 (1988): 481–511, 736–64. Maddy offers a detailed discussion about results on large cardinals, in the wake of Gödel's 1947 article.

21 Apart from ideas contained in the book on *"Mathematische Existenz,"* already mentioned, Becker established interesting links between phenomenology and the foundations of geometry. See Becker, *"Beiträge zur phänomenologischen Begründung der Geometrie und ihrer physicalischen Anwendungen," Jahrbuch für Philosophie une phänomenologische Forschung* 6 (1923): 385–560. For the notion of *strenge Implikation*, which led him to the development of one of the first systems of modal logic, see Becker, "Zur Logik der Modalitäten," *Jahrbuch für Philosophie une phänomenologische Forschung* 11 (1930). On the issue of modality, see Weyl's different approach. Weyl,

"The Ghost of Modality," in *Philosophical Essays in Memory of Edmund Husserl*, ed. Marvin Farber, 278–303 (Cambridge MA: Harvard University Press, 1940). None of these important developments of the phenomenological standpoint and their relation to intuitionism can be discussed here.

22 It is probable that Brouwer attended Husserl's lecture in Amsterdam in 1928. They met and Husserl described meeting Brouwer in very enthusiastic terms in a letter to Heidegger, but nothing concrete seems to have come out of this meeting. Husserl, *Briefwechsel*, 10 vols. (Dordrecht: Kluwer, 1994), 4: 156.

23 I must also leave aside the recent, very interesting attempt by Mark van Atten in his *Phenomenology of Choice Sequences*, to argue that phenomenology can and should provide grounding for Brouwer's theory of choice sequences.

24 See Roman Ingarden, *On the Motives which led Husserl to Transcendental Idealism* (The Hague: Martinus Nijhoff, 1975). For a defense of Ingarden's arguments see Karl Ameriks, "Husserl's Realism," *Philosophical Review* 86 (1977): 498–519. Ingarden was far from being the only one to have disagreed, at the time, with Husserl's transcendental turn: the whole of the Münich School (e.g., Johannes Dauben, Adolf Reinach, etc.), and almost all the Göttingen students (such as Ingarden), for example, very much appreciated *Logical Investigations* and rejected *Ideas* I and its Kantian transcendentalism. Heidegger also disliked *Ideas* I; it is quite clear that he held Husserl's notion of categorial intuition, as introduced in chapter six of the sixth of the *Logical Investigations*, as a decisive breakthrough; he saw it as opening up the *Seinsfrage* with, in a nutshell, 'being' or 'Being' as the objective correlate of 'to be.' See Martin Heidegger, *Zur Sache des Denkens* (Tübigen: Niemeyer, 1969), 86 and *Vier Seminare* (Frankfurt am Main: Klostermann, 1977), 116. Heidegger clearly latched onto remarks such as this one from § 43: "I can see colour but not *being*-coloured," Husserl, *Logical Investigations*, II, 781. This is why Husserl proposed categorial intuition as a 'widening' of the concept of perception, to account for the grasping of, *inter alia*, 'Being' as the objectivity correlated with the copula.

25 Those who defend the idea that Husserl was a (transcendental) idealist through and through and that attempts at interpreting his earlier position, at the time of *Logical Investigations*, as a form of realism distort the true import of his phenomenology usually quote the following: "Only someone who misunderstands either the deepest sense of intentional method, or that of transcendental reduction, or perhaps both, can attempt to separate phenomenology from transcendental idealism." Husserl, *Cartesian Medita-*

tions (The Hague: Martinus Nijhoff, 1960), 86. This passage merely shows, however, that Husserl had indeed changed his mind and it is not *per se* an argument. Moreover, in a later letter to Émile Baudin in 1934, Husserl wrote he was not anymore using the word idealist to describe his philosophy. Husserl, *Briefwechsel*, 7: 16. I cannot even begin to discuss here this vast topic; I should merely state that it is rather attempts at reading the later phenomenology into *Logical Investigations* that create distortions and make for a shallow understanding of that great work. I should add also that I am in agreement with Findlay, when he wrote that "Those who, like most of Husserl's French interpreters, have approached him from an exclusively subjectivistic, neo-Kantian standpoint, basing themselves on writings since 1913, and who largely bypass or ignore Brentano, never will rise to a full understanding of his phenomenology." John N. Findlay, "Phenomenology and the Meaning of Realism," in *Philosophy and Phenomenological Understanding*, ed. Edo Pivcevic, 143–58 (Cambridge: Cambridge University Press, 1978), 146.

26 Dagfinn Føllesdal, "Gödel and Husserl," in *From Dedekind to Gödel: Essays on the Development of the Foundations of Mathematics*, ed. Jaakko Hintikka, 427–446 (Dordrecht: Kluwer, 1995). See also Wang, *Reflections*, 121.

27 Husserl, *Briefwechsel*, 7: 287.

28 The presentation in the following paragraphs is partly based on that in Guillermo E. Rosado Haddock, "Husserl's Epistemology of Mathematics and the Foundation of Platonism in Mathematics," *Husserl Studies* 4 (1987): 81–102.

29 As Husserl called an 'essence' an '*eidos*,' one may also speak of 'eidetic' intuition. For further discussions of the *Wesensschau* by Husserl, see *Ideas* I, part I and *Phenomenological Psychology: Lectures, Summer Semester, 1925* (The Hague: Martinus Nijhoff, 1977), 53–65.

30 Husserl, *Experience and Judgement* (Evanston IL: Northwestern University Press, 1973).

31 Husserl, *Experience and Judgement*, §§ 58ff.

32 See Haddock, 88–9, who speaks of a radicalization of the contrast, from *Logical Investigations* to *Experience and Judgement*.

33 Translation modified. In the introduction to *Space-Time-Matter*, Weyl also presents an idealistic, Kantian version of phenomenology. See Weyl, *Space-Time-Matter* (New York: Dover, 1952).

34 See Da Silva, "Husserl's Phenomenology and Weyl's 'Predicativism,'" *Synthese* 110 (1977): 277–96 on Fichtean elements in *The Continuum* and John L. Bell for a discussion of "Insight and Reflection" in "Hermann Weyl

and *The Continuum," Philosophica Mathematica* 3 (8) (2000): 259–273.

35 Again, I cannot, within the scope of this paper, cover in sufficient details all the issues involved here. The points I am making are in agreement with the more detailed analyses in Bell and Da Silva.

36 Weyl, *The Continuum*, 19.

37 See section 51, *in finem*, of the sixth of the *Logical Investigations*. On this point, see Kaufmann, chapter one, on the "Basic Facts of Cognition." To summarize in a very compressed manner: according to Kaufmann, every *bona fide* mathematical statement must be reducible to statements about natural numbers, themselves understood as "logical abstractions of the process of counting," a claim which he expresses by saying that "every particular natural number is a last logical specification, that is a formal eidetic singularity in Husserl's sense." Kaufmann saw this requisite as allowing only for talk of "properties of material objects" and blocking the way to talk of "properties of properties", "properties of properties of properties," and so forth, thus to transfinite set theory, because such talk cannot be reduced to talk about natural numbers and thus no verification can be made on the basis of founding acts. This verificationist component is also present in Becker and Heyting. Kauffman, *The Infinite in Mathematics*, 22–3, 71.

38 See Henri Poincaré, *Science and Hypothesis* (New York: Dover, 1952), 12–13.

39 On this point see Da Silva, "Husserl's Phenomenology," 292 and Tieszen, "Phenomenology," 394ff.

40 See the articles reproduced in Gerhard Heinzmann, *Poincaré, Russell, Zermelo et Peano: Textes de la discussion (1906–1912) sur les fondements des mathématiques: des antinomies à la prédicativité* (Paris: Albert Blanchard, 1986).

41 Husserl, *Phenomenological Psychology*, 53.

42 This argument is spelled out in much detail in Da Silva's "Husserl's Phenomenology."

43 Da Silva's "Husserl's Phenomenology," 279–80.

44 Weyl, *The Continuum*, 90.

45 Weyl, *The Continuum*, 93.

46 On this point, see Bell.

47 Da Silva, "Husserl's Phenomenology," 289.

48 Da Silva, "Husserl's Phenomenology," 293.

49 Hao Wang, *From Mathematics to Philosophy* (New York: Humanities Press, 1974), 8–10.

50 Gödel, *Collected Papers III*, 320.

51 Gödel, *Collected Papers II*, 268.

52 Gödel, *Collected Papers II*, 268.
53 Gödel, *Collected Papers III*, 359.
54 Gödel, *Collected Papers III*, 305.
55 Gödel, *Collected Papers III*, 383.
56 Føllesdal, "Godel and Husserl," 428.
57 See Tieszen's papers: "Gödel's Path from the Incompleteness Theorems (1931) to Phenomenology (1961)," *The Bulletin of Symbolic Logic* 4 (1998): 181–203 and "Kurt Gödel and Phenomenology," *Philosophy of Science* 59 (1992): 176–94.
58 I should like to thank Richard Feist, Jairo José Da Silva, and Hermann Philipse for insightful discussions on the content of this paper, even though I did not always agree with their views. I should like also especially to thank my student Anoop Gupta, with whom I originally planned to write this paper.

CHAPTER SEVEN

Richard Feist

HUSSERL AND WEYL: PHENOMENOLOGY, MATHEMATICS, AND PHYSICS

―――

1. INTRODUCTION

In the early years of the twentieth century, Edmund Husserl and Hermann Weyl exchanged a number of letters on the foundations of mathematics and science.[1] On 10 April 1918, Husserl writes that reading Weyl's *The Continuum* was a meaningful event since he himself had been on a "similar path" for many years.[2] The following comment of Husserl's offers some idea as to what he meant by this similar path.

> At last a mathematician, who, by understanding the necessity of a phenomenological approach to all questions concerning the clarification of foundational concepts, returns to the primary ground of logico-mathematical intuition, the sole basis upon which a real foundation of mathematics and an insight into the sense of mathematical results is possible![3]

According to Husserl, then, Weyl's path is a systematic, phenomenological approach to the foundations of knowledge based on a logico-mathematical intuition. Husserl sees Weyl's work in physics as on the same path. On 5 June 1920, Husserl applauds Weyl's study of General Relativity, *Space-Time-Matter*, stating that it is

―――

close to his own "ideal of a physics permeated by philosophical spirit."[4] Husserl expresses his fascination for Weyl's "peculiar and deep" knowledge of Riemannian space and that:

> as an ex-mathematician, I presume to understand the *sense* of the deductions. In *my* intellectual perspective, I am always concerned with transcendental meaning, which leads to similar, correlated problems. That is why I am drawn to theories such as yours.[5]

Perhaps Husserl's remarks are simply an attempt to garner support for his philosophical programme. For there had been attacks on phenomenology by 'scientifically'-minded philosophers, such as Moritz Schlick, who specifically criticized Husserl's idea of a non-sensuous intuition.[6] Schlick had pejoratively referred to Husserl and Henri Bergson as the "prophets of intuition." So what better response could there be than to claim that Weyl, perhaps second only to Einstein in prestige at the time, was a follower of the phenomenological method?

There is probably some truth to this biographical explanation of Husserl's remarks. Indeed, Husserl's response to Schlick was more anger than argument.[7] But this explanation obscures the philosophical connections between Husserl's and Weyl's work. Moreover, Weyl responded to Schlick's criticisms of Husserl in a review of Schlick's *General Theory of Knowledge.* Although Weyl sometimes expresses great respect for Schlick, he nonetheless insists that anyone who has had the phenomenological experiences that Husserl describes, the non-sensuous intuitions involved in the foundations of logic and mathematics, cannot possibly accept Schlick's epistemology, which errs by limiting intuition to the "merely sensuous."[8]

Unfortunately, Weyl neither explains these "phenomenological experiences" at the basis of logic and mathematics nor explicitly claims to have had them himself. However, I hold that Weyl's logico-mathematical intuition can be understood as though it is a Husserlian categorial intuition.[9] To support this claim I shall read Weyl's *The Continuum* and *Space-Time-Matter* as mutually supplementary texts. Nonetheless, we shall see that it is difficult to catch Weyl's thought with a single net since it is not free of ambiguities.

I will then discuss how Weyl's work in the foundations of Riemannian geometry, which lies at the basis of Relativity Theory, serves to alter the relation between Weyl's thought and phenomenology. Here, too, ambiguities arise.

2. THE SYSTEMATIC NATURE OF WEYL'S THOUGHT

Weyl claims in *Space-Time-Matter* that he will present General Relativity systematically, as an "illustration of the intermingling of philosophical, mathematical, and physical thought."[10] This intermingling is the most pronounced, he continues, in the problem of space. In an early edition of *Space-Time-Matter*, Weyl states that mathematicians should be proud of the representative breadth of mathematics, but humble since mathematics (i.e. geometry) cannot say anything about space that it does not also say about states of addition machines or systems of linear equations. Mathematics, then, only offers the most superficial and formal aspects of space.[11] It is metaphysics, Weyl says, that tells us what is unique to space. In general, metaphysics or philosophy – Weyl often uses them interchangeably – provides the experiential origins of all our scientific concepts. Moreover, the knowledge provided by metaphysics is crucial. Weyl writes:

> All origins [of concepts] are unclear. Inasmuch as the mathematician operates with concepts in a strict and formal fashion in constructing science, he must be reminded from time to time that origins in their dim and distant depths refuse to be grasped by his methods. Beyond the knowledge gained from the individual sciences, there remains the task of *comprehending*. Although philosophy discouragingly swings to and fro, from system to system, we cannot dispense with it lest we transform knowledge into a meaningless chaos.[12]

Since this remark occurs in all editions (thus surviving all revisions) of *Space-Time-Matter*, it would *seem* that for the bulk of Weyl's career the systematic nature of philosophy remained an important element of his thought. My immediate concern, however, is the nature of Weyl's thought around the time of his correspondence with Husserl.

We can acquire an idea of what this system is by considering Weyl's discussion of geometry's history. He claims that Euclid's early commentators first opened the possibility of non-Euclidean geometry when they realized that the Fifth postulate was not self-evident and then attempted to prove it on the basis of those that were. Weyl continues:

> A report of Proclus (A.D. 5) about these attempts has been handed down to posterity. Proclus utters an emphatic warning against the abuse that may be practised by calling propositions self-evident. This warning cannot be repeated too often; on the other hand, we must not fail to emphasize the fact that, in spite of the frequency with which this property is wrongfully used, the "self-evident" property is the final root of all knowledge, including empirical knowledge.[13]

So the task is to clarify 'self-evidence.'[14] To begin, Weyl often claims to follow Husserl's philosophy. In *The Continuum* he states that he agrees with Husserl's epistemology of logic in *Logical Investigations* and Husserl's attempt to embed this epistemology of logic within a broader, systematic philosophy, the philosophical system articulated in *Ideas*.[15] Moreover, in *Space-Time-Matter* he again claims to be following Husserl's philosophy as presented in *Ideas*.[16] But as we shall see, it is not entirely clear as to what aspects of Husserl's thought Weyl is committing himself.

Self-evidence plays a key role in Weyl's analogy between the foundations of empirical science and logic. Empirical science's most fundamental idea is the externality of the world, the "thesis of reality."[17] Logic's most fundamental idea is the objectivity of truth, the "thesis of truth."[18] Explicating the meaning and justifying the assertion of these theses can only be done on the basis of that which is 'self evident' or 'absolutely given' or 'immediately given.'[19] In other words, the foundations of the sciences and of mathematics are given after the phenomenological reduction since it is only then that consciousness is presented with that which is 'immediately given.'[20] But it is still unclear as to what this 'immediately given' is.

In *The Continuum* Weyl claims that the given in intuition is, "first and foremost, a fluid whole rather than a set of discrete elements."[21] Let us call this the 'Base Intuition,' whose correlate is a 'fluid whole.' Now the Base Intuition does *not* provide consciousness with 'objects' in the ordinary sense of the term. The Base Intuition is a more elementary experience than the experience of objects. In *Space-Time-Matter*, Weyl tells us more about the Base Intuition: it is prior to perception and is essentially an extremely vague feeling of a causal interaction with the world.[22] The Base Intuition then only serves to demarcate the subject (the individual consciousness) from the object (the fluid whole or the continuous world).[23] Since the Base Intuition is essentially a causal feeling, a direct link to the world, consciousness and the world are not radically separated. So it is in this sense that Weyl says that consciousness and the world are inextricably bound.[24]

Perception, then, begins 'after' the felt causation. Here we are given objects, although they are given via profiles.[25] Each profile, Weyl says, is immediately given; it can, along with the experiential act that 'contains' it, be raised by free acts of reflection so that its essence can be brought to intuition.[26] In sum, Weyl is *briefly* describing how, under the phenomenological reduction, essences can be brought to intuition. At first, Weyl is concerned with bringing forth the general essence of the concept 'real object.' By this he means 'empirical object' since he holds that ideal objects are also real (mind-independent). Weyl writes:

It lies in the essence of a real thing to be to be inexhaustible in content; we can get an ever deeper insight into this content by the

continual addition of new experiences, partly in the apparent contradiction, by bringing them into harmony with one another. In this
sense the real thing is a limiting idea. From this arises the empirical
character of all our knowledge of reality.[27]

So the *essence* of 'empirical object,' once brought to intuition,
reveals that the entities that fall under this essence – the empirical objects themselves – have inexhaustible contents. Weyl, like
Husserl, leaves unanswered just how we 'intuit' this inexhaustibility. In *Ideas* Husserl claims that the empirical object is a Kantian
Idea.[28] Indeed, Weyl agrees:

> A real thing can never be given adequately, its "inner horizon" is
> unfolded by an infinitely continued process of ever new and more
> exact experiences; it is, as emphasized by Husserl, a limiting idea in
> the Kantian sense. For this reason it is impossible to posit the real
> thing as existing, closed and complete in itself. The continuum
> problem thus drives one toward epistemological idealism.[29]

In sum, the examination of perception, which was under the
phenomenological reduction and so bracketed the 'thesis of reality,' has demarcated the domain of empirical objects, namely,
those objects that are presented in profiles. For Weyl, like Husserl,
because we can bring the essence of 'empirical object' to intuition, we are able to say that the domain of empirical objects is
extensionally determinate. That is, they pre-exist and so what
can possibly count as an empirical object is well-defined.

Another way to understand this is that 'empirical object' has
a determinate extension; all objects within this extension are of
the same ontological order. The boundaries of this extension,
then, are stable.

Now let us turn to mathematics. Here the situation is much
trickier. Weyl flatly states that the universal concept, 'object,' is
not extensionally definite.[30] For here we can construct objects of
arbitrary (ideal) orders. 'Object,' then, has an unstable boundary.
Since 'empirical object' is extensionally definite, Weyl's concern

is that 'ideal object' is not extensionally definite. This is a serious problem since Weyl insists that mathematics is based on a certain kind of ideal object, namely, the set of natural numbers. Mathematics then proceeds to *construct* its other ideal objects on the basis of the natural numbers.[31] Weyl imposes restrictions on the acceptable constructions of ideal objects, restrictions which led to the rejection of the least upper bound theorem (for bounded sets of reals, but not series of reals) and Dirichlet's principle. However, I am not going to enter that issue here.[32] What I am going to investigate is simply how the natural numbers are given to us according to Weyl.

To begin, Weyl insists that we do not directly intuit the numbers.[33] He states that the numbers form a category of ideal objects and that there is a single basic relation that underlies this category, namely, the successor relation.[34] Now Weyl claims that we intuit the *meaning* of this relation. That is, the meaning of the successor relation is 'immediately intuited.'[35] In other words, it would seem that we are able to intuit the essence of a *particular* ideal object category, 'natural number.'

Further support for the categorial reading can be found in Weyl's discussion of the concept of existence. He states that if a statement such as '∃xPx is true' is meaningful, then it asserts a definite state of affairs and that the statement's truth value can be investigated. The meaningfulness of the statement illustrates the determinacy of the quantifier's domain. Weyl states that this determinacy is a result of the characteristics pertaining to the "categorical essence under consideration."[36] Since he applies these remarks regarding meaning to the domain of natural numbers, it would be safe to conclude that the *meaning* which we intuit to underlie the natural numbers, that is, the category 'natural number,' is some kind of categorial essence.

Claims like this regarding the category of 'natural number' sound very much like those we have seen Weyl make regarding the category 'empirical object.' Now the intuition of 'empirical object,' as we have seen, is achieved after acts of reflection upon the perceptually given – which is ultimately founded upon the correlate of the Base Intuition. We also saw that the intuition of 'inexhaustibility' came later in the series of reflections. So, if we take Weyl seriously that the Base Intuition does present the 'whole primary story,' then it would seem that the intuition of the essence 'natural number' is also some kind of a founded intuition.

Unfortunately, things are not always so straightforward when reading Weyl's texts. Weyl sometimes speaks as though the basic intuition in mathematics is really an intuition of iteration.[37] So in this case, it is not the intuition of a categorial essence of a set of entities. Rather, he simply means that we have some kind of an *unfounded* intuition of iteration, the repeatability of a discrete operation.[38] Regarding the intuition of iteration, Weyl states that he is in agreement with Poincaré.[39]

But there still remains a difficulty to face. As we have seen, Weyl claims to agree with Husserl's epistemology of logic in *Logical Investigations*. But this text itself contains numerous discussions of the foundations of logic, and Husserl's adherence to parts of it (especially that of categorial intuition) wavered around the time that Weyl was writing *The Continuum* and *Space-Time-Matter*.[40] But, in *Prolegomena*, Husserl holds that logical truths are simply given in some kind of unfounded intuition, which also includes Bernoullian induction, the inference from n to n+1.[41] So it might be the case that when Weyl agrees with Husserl's epistemology, he is agreeing with that of *Prolegomena* only.

The problem, as I see it, is this. We could read Weyl's fundamental intuition in mathematics either as a categorial style intuition or as an unfounded immediate intuition of iteration. We could find passages from *The Continuum* to support either reading. Admittedly, *The Continuum* leans toward an immediate unfounded reading. But I maintain that if we read *Space-Time-Matter* in conjunction with *The Continuum*, and take Weyl seriously in his claims regarding the importance of systematic philosophy and the primary nature of the correlate of what I have called the Base Intuition, then we should lean towards a categorial-style reading of the fundamental intuition that grounds mathematics.

3. RIEMANNIAN MANIFOLDS AND PHENOMENOLOGY

Soon after the first edition of *Space-Time-Matter* appeared Weyl modified his view regarding the superficiality of mathematics with regards to the essence of space. That is, he held that geometry had undergone a radical transformation such that it was able to disclose the essence of space and was no longer a merely formal enterprise. Mathematics, then, had stepped onto the terrain previously occupied by metaphysics. Nonetheless, Weyl did not think that mathematics had displaced metaphysics as a viable enterprise. Rather, he thought that the deepening of the representative power of mathematics served to bring the disciplines of mathematics and phenomenology closer together. One could say that Weyl held that mathematics could serve as a *corrective* to phenomenological investigations. Mathematics helps to reveal more clearly that which is truly presented in consciousness.

Interestingly enough, some years before Weyl worked in the foundations of infinitesimal geometry, Husserl suggested in *Ideas* that there might be some kind of symbiotic relationship between geometry and phenomenology. Although Husserl insists that phenomenology and mathematics are distinct disciplines, he investigates whether they are analogous. After articulating the concept of a 'definite manifold' that is, a region or province of being such that from a finite number of concepts and axioms, all possibilities within that province can be derived, Husserl asks: can the stream of consciousness – the so-called manifold of experience – be regarded as a mathematical manifold?[42] Another way Husserl puts the question is: could phenomenology be understood as a 'geometry of experience?' Husserl states that it cannot. There is a necessary condition for a concrete manifold to satisfy in order to be brought under the idea of a definite manifold, namely, that the manifold in question consists of *exact* essences. But the stream of experience does not give exact essences; rather, it gives morphological essences, which have vague boundaries and are therefore nonmathematical.[43] The main point is that Husserl insists that the stream of consciousness is not a mathematical manifold and so geometry and phenomenology are distinct.

Phenomenologists, however, may mimic the geometers by constructing laws that concatenate morphological essences in an effort, so to speak, to 'axiomatize experience.' But, Husserl insists that:

> an actual seeing of the concatenations of essences must redeem the presumed likelihoods. As long as that has not occurred, we have no phenomenological result.[44]

Nonetheless, Husserl mitigates the opposition between the two disciplines by claiming that his entire discussion of the difference between geometry and phenomenology:

> does not answer the pressing question of whether, *besides* the descriptive procedure, one might not follow – as a counterpart to *descriptive* phenomenology – an idealising procedure which substitutes pure and strict ideals for intuited data and might even serve as the fundamental means for a mathesis of mental processes.[45]

In sum, Husserl admits that mathematics could, in some sense, serve to assist phenomenology.

When *Space-Time-Matter* first appeared in 1918, Weyl held that an important epistemological view, adumbrated decades earlier by Bernhard Riemann, had finally taken hold in physics via General Relativity.[46] Weyl wrote:

> To the transition from Euclid's distant-geometry to Riemann's contact-geometry corresponds the transition from action-at-a-distance physics to action-by-contact physics ... The principle that the world should be understood through its behaviour in the infinitely small is the leading epistemological force behind both the physics of action by contact and Riemann's geometry.[47]

However, later that same year Weyl developed Levi–Civita's work concerning parallel displacements on Riemannian manifolds and recanted the above passage. Intuitively, parallel displacement is unproblematic; it preserves magnitude and direction in a path-independent manner. But path independence holds only for length on Riemannian manifolds. Because of the non-integrability of direction, some mathematicians declared that parallel displacement is meaningless on Riemannian manifolds. Levi–Civita thought differently, namely, that parallel-displacement is meaningless on Riemannian manifolds *provided* that the displacement is finite. In sum, parallel displacement makes perfect sense on Riemannian manifolds in the case of infinitesimal displacement.

The crux of the issue is the description of these infinitesimal displacements on Riemannian manifolds. In order to describe such displacements, Levi–Civita embedded the manifold in a higher-order space. Weyl, however, demonstrated that one could describe the infinitesimal displacements intrinsically, obviating the need for a higher-order embedding space. Now, it is important to recall Riemann's epistemological principle of understanding the world through its behaviour in infinitesimal regions. Because vector direction is non-invariant under finite parallel transport, and so must be understood infinitesimally, Riemann's principle would demand no less with respect to length. That is, length invariance under finitistic parallel transport cannot be assumed. Riemann never asked this question; he simply took the path-independent nature of length for granted. In the fourth edition of *Space-Time-Matter* Weyl writes:

> Inspired by the weighty inferences of Einstein's theory to examine the mathematical foundations anew the present writer made the discovery that Riemann's geometry only goes half-way towards attaining the ideal of a pure infinitesimal geometry. It still remains to eradicate the last element of geometry 'at a distance,' a remnant of its Euclidean past. Riemann assumes that it is possible to compare the lengths of two line elements at *different* points of space, too; *it is not permissible to use comparisons at a distance in an 'infinitely near' geometry.* One principle alone is allowable; by this a division of length is transferable from point to that infinitely adjacent to it.[48]

Weyl had stated in an earlier edition of *Space-Time-Matter* that geometry was a superficial study of space, saying no more about space than it said about other complex objects such as systems of linear equations. Geometry, now understood as *infinitesimal geometry*, is what Weyl calls a 'true geometry,' that is:

> a doctrine of *space itself* and not merely like Euclid, and almost everything else that has been done under the name of geometry, a doctrine of the configurations that are possible in space.[49]

Weyl's recasting of the General Theory of Relativity within a modified geometry enabled the theory to account not only for gravitation, but for electromagnetism as well. This was indeed a major achievement, for at that time gravitation and electromagnetism were the only known forces in the universe. No wonder Weyl thought that the dream of a unified theory had finally been fulfilled.[50]

In the conclusion of his discussions on the 'new' mathematics of space, Weyl writes:

> The investigations about space that have been conducted in Chapter II seemed to the author to offer a good example of the essential analysis [*Wesenanalyse*] which is the object of Husserl's phenomenological philosophy – an example which is typical of cases in which we are concerned with nonimmanent essences.[51]

Nonimmanent essences, that is, the essences of external objects, are extremely difficult for the mind to grasp. Historical prejudices hinder (not necessarily prevent) the mind from seeing what truly lies before it. What Weyl is drawing attention to here is that it took considerable time to completely overcome the Euclidean prejudice of permitting 'comparison-at-a-distance.'

Nonetheless, Weyl admits that *in principle* the philosophers are correct to say that the mind can grasp the essences of external objects on the basis of a single intuition. But *in practice* they are mistaken. Weyl concludes:

The problem of space is at the same time a very instructive example of that question of phenomenology that seems to the author to be of greatest consequence, namely, how far the delimitation of the essentialities perceptible in consciousness expresses the structure peculiar to the realm of presented objects, and in how far mere convention participates in this delimitation.[52]

Mathematics assists phenomenological investigations of experience by eradicating conventions so that the essential structures of the given more fully present themselves. So Weyl, like Husserl, holds that mathematics can serve as a corrective to phenomenological investigations.

However, to read Weyl as regarding mathematics as a *corrective* to phenomenology is not to imply that he subordinated mathematics to phenomenology. Consider Weyl's 26 March 1921 letter to Husserl:

> During recent times I managed to understand the essence of space as far as the deepest foundations which can be reached by mathematical analysis. Therefore one faces research on group theory like that, which Helmholtz in his time, and you have carried out on the problem of space. However, today with the theory of relativity one has to consider that the situation has changed ... One has to sever the *a priori* essence of space expressed in the metric from the *a posteriori* of the orientation of this metric at different points, which is determined in nature by matter.[53]

There are, so to speak, two *a priori* essences of space. Moreover, because of developments in mathematics and science they have finally been separated. The first *a priori* essence of space is a phenomenological *a priori*; it is what Weyl calls the "the metric of space." This states that at each point in space and its infinitesimal neighbourhood, the space is Euclidean. Here mathematics and phenomenology coincide in their results. Phenomenologically, space is revealed to us as Euclidean and Weyl thought that this Euclidean nature could be mathematically proven to hold in the *infinitesimal* domain.[54] Here we have Weyl's version of Husserl's 'side-by-side' relationship between mathematics and phenomenology.

Weyl later offered a metaphor to explain the relationship between phenomenology and science with regard to space. Throughout his life Weyl maintained that we have an intuition of space and that these are 'Euclidean intuitions.' He held that these intuitions do not conflict with science provided that we restrict this intuition to infinitesimally small neighborhoods of a point O. Weyl continues:

> But then one has to allow that the connection between intuitive and physical space becomes increasingly vague the further one departs from O. This is analogous to a tangential plane (intuitive space) touching a point O of a curved surface (physical space).[55]

For Weyl, there is no doubt that General Relativity has revealed that real space is not globally Euclidean. There is a fairly straightforward realism in Weyl's thinking on this issue.[56] Let us consider another of Weyl's comments to Husserl.

> Outside the phenomenological *a priori* there also exists a metaphysical one, a necessity, which does not appear clear from the intuition of the relative essence, but is only revealed from its metaphysical meaning. However, to understand this meaning, it seems to me, that also the knowledge comes from experience plays a principle indispensable role.[57]

Ultimately, there is an ambiguity in Weyl's writings as to the intuition of space. In *Space-Time-Matter* he seems to lean towards the view that it is only the accidental aspects of the human condition that prevent consciousness from directly intuiting the structure of space. Recall that he says that, *in principle* consciousness can do this, but from the point of view of human nature (the concrete human condition), it cannot. Moreover, with respect to the historical development of science and its supposed liberation of consciousness from the accidental prejudices regarding the essence of space, Weyl states that "reason" (consciousness) is somehow always aware of the "correct point of view." After the scientific development, reason is "flooded with light" and comprehends that which is of "itself intelligible to it." But again, reason alone lacks the power to penetrate into the correct point of view simply in one "flash."[58] In his correspondence with Husserl, Weyl seems to hold a stronger position in that the activity of science truly does reveal something about the world that cannot, even in principle, be revealed in any other way.

But there is a common element to these ambiguities in Weyl's philosophy. In both cases there is a tension between Platonistic and constructivistic tendencies. This is reflected in Weyl's partial allegiance to founding intuitions and constructions or activities based on these intuitions. This commonality can be located within the greater context of Weyl's intellectual career. Indeed, as I see it, his entire career could be described as a slow movement from Platonism to constructivism. In his early years he is a strong Cantorian[59] and one can see his ontological commitments shrink over time.[60]

4. FINAL COMMENTS

I am certainly not the first to notice ambiguities in Weyl's thought. Stephen Toulmin asserted (in his review of Weyl's *Philosophy of Mathematics and Natural Science*) that Weyl performs more the duties of a "godfather than legitimate parent" to the philosophical ideas in his care. Moreover, Weyl's discussions have a "decidedly warmed up flavour: everything comes out in other people's technical terms."[61] Indeed, the cobbled feel of the text under review is hardly surprising, given Weyl's comment that it was written after he had spent "a satisfying year of reading philosophy – as a butterfly darts from one blossom to the next, troubled, yet drawing honey from each."[62] However, Weyl generalizes this "butterfly approach" to his other works when he writes:

My own mathematical works were always quite unsystematic, without method or connection. Expression and shape are almost more important to me than knowledge itself.[63]

So with comments like this from Weyl himself, it is not at all surprising that interpretative difficulties arise since different philosophical positions can be extracted from his writings, especially those of *The Continuum* and *Space-Time-Matter*. As we have seen, Weyl at one time in his career stressed that knowledge *must* have a systematic philosophical context lest it "slide into chaos." However, he also admitted that he drew upon various sources of philosophy in what appears to be an unsystematic manner. But does this mean that Weyl ultimately did not take philosophy itself seriously? Is Toulmin correct when he says that, for Weyl, philosophizing was, at most, a "part-time activity?"[64] I would suggest not. Simply because we can find ambiguities in Weyl's works that does not mean that he did not take the philosophical issues seriously. I submit that what we should infer from all this is that Weyl's "philosophical spirit" was indeed a restless one. For like a butterfly, it drank heavily from that flower upon which it landed, although it always had one eye turned to the next.

Notes

1 These letters can be found in Dirk van Dalen, "Four Letters from Edmund Husserl to Hermann Weyl," *Husserl Studies* I (1984): 1–12.

2 Van Dalen, "Four Letters from Edmund Husserl to Hermann Weyl," 3.

3 Van Dalen, "Four Letters from Edmund Husserl to Hermann Weyl," 3.

4 Van Dalen, "Four Letters from Edmund Husserl to Hermann Weyl," 5.

5 Van Dalen, "Four Letters from Edmund Husserl to Hermann Weyl," 5.

6 Moritz Schlick, *Allgemeine Erkenntnislehre,* (Berlin: Springer Verlag, 1918), 120–21.

7 Husserl's response to Schlick is in the second edition of *Logical Investigations* published in 1921. This response to Schlick can be found in the Findlay translation (London: Routledge and Kegan Paul, 1970), 663–64.

8 Weyl, "Review of Schlick's *Allgemeine Erkenntnislehre,*" *Jahrbuch über die Fortschritte der Mathematik* 46 (1923): 59.

9 For a different view of Weyl's logico-mathematical intuition, see Marion's paper in this volume.

10 Weyl, *Space-Time-Matter*, 4th ed., tr. H.L. Browse (New York: Dover, 1952), ix.

11 Weyl, *Space-Time-Matter*, 26.

12 Weyl, *Space-Time-Matter*, 10.

13 Weyl, *Space-Time-Matter*, 77.

14 Although some commentators may differ with my reading of Weyl on 'self-evidence' and its relation to mathematics, I am trying to support the claim made some time ago by Peter Beisswanger, who says that Weyl "had in fact employed Husserl's concept of *Evidenz* [*Evidenzbegriffe*] in order to free number theory from the hand of set theory." "Die Phasen in Hermann Weyls Beurteilung der Mathematik," *Mathematisch-Physikalische Semesterberichte*, New Series 12 (1965): 139.

15 Weyl, *The Continuum: A Critical Examination of the Foundations of Analysis* (NY: Dover Publications, 1994), 3.

16 Weyl, *Space-Time-Matter*, 319.

17 Weyl, *Space-Time-Matter*, 4.

18 Weyl, *Space-Time-Matter*, 5.

19 Weyl, *Space-Time-Matter*, 4–5.

20 For Weyl's brief description of the phenomenological reduction, see *Space-Time-Matter*, 4.

21 Weyl, *The Continuum*, 49.

22 Weyl, *Space-Time-Matter*, 6.

23 The notion of 'felt causation' was certainly in the academic air of the time. It has its partial roots in Fichte's thought; however, to say anything more regarding the Weyl–Fichte connection would involve more on Weyl's connection to Husserl since the latter was also working on Fichte at this time. Moreover, 'felt causation' also plays an important role in Henri Bergson's thought. Weyl praises Bergson's important contributions to our understanding of the relationships between mathematical concepts and the nature of our experience. Weyl, *The Continuum*, 90. The view that the primordial contact with the world is a 'causal feeling' is the foundation of Whitehead's metaphysics, which was under construction at this time. These, however, are stories for another time.

24 Weyl, *Space-Time-Matter*, 6.

25 Weyl, *Space-Time-Matter*, 4.

26 Weyl, *Space-Time-Matter*, 4.

27 Weyl, *Space-Time-Matter*, 5.

28 Husserl, *Ideas Pertaining to a Pure Phenomenology and to a Phenomenological*

Philosophy: First Book, tr. F. Kersten (The Hague: Martinus Nijhoff, 1983), § 143.

29   Weyl, *Philosophy of Mathematics and Natural Science*, 41.

30   Weyl, *The Continuum*, 110.

31   Weyl, *The Continuum*, 25.

32   For discussions along these lines see Marion's paper in this volume as well as Dirk van Dalen, "Hermann Weyl's Intuitionistic Mathematics," *The Bulletin of Symbolic Logic*, 1 (1995): 145–169. But suffice it to say that according to Weyl, introducing concepts with flexible boundaries into mathematics is the source of the paradoxes in the foundations of set theory. See Weyl's discussion of Russell's paradox (in the form of Grelling's) on page 6 of *The Continuum*. Finally, see R. Feist, "Weyl's Appropriation of Husserl's and Poincaré's Thought," *Synthese* 132 (3) (2002): 273–301.

33   Weyl, *The Continuum*, 15. Weyl seems to hold the view that a natural number is really a universal. This view is also held by Husserl in *Logical Investigations* and B. Russell in *The Problems of Philosophy* (Oxford: Oxford University Press, 1912), chapter 10. For more on the relations between Husserl and Russell see Dallas Willard, *Logic and the Objectivity of Knowledge: A Study in Husserl's Early Philosophy* (Ohio: Ohio University Press, 1984).

34   Weyl, *The Continuum*, 25.

35   Weyl, *The Continuum*, 25.

36   Weyl, *The Continuum*, 8.

37   Weyl, *The Continuum*, 48.

38   Weyl, *The Continuum*, 19.

39   Weyl, *The Continuum*, 48. However, immediately after expressing agreement with Poincaré, Weyl distances himself from Poincaré's general thought.

40   See Husserl's comments that show that he distanced himself from the original presentation of 'categorial intuition,' preface to the second edition of *Logical Investigations*, 45.

41   Husserl, *Logical Investigations*, 99.

42   Husserl, *Ideas*, 161.

43   Husserl, *Ideas*, 162. Weyl also uses Husserl's language of morphological essences to describe what is given within the Base Intuition. *The Continuum*, 49.

44   Husserl, *Ideas*, 169.

45   Husserl, *Ideas*, 169.

46   For a more involved, yet straightforward discussion of the mathematics, see Alberto Coffa, "Elective Affinities: Weyl and Reichenbach," in *Hans Reichenbach: Logical Empiricist*, ed. Wesley Salmon, 267–304 (Holland: D.

Reidel Publishing Company, 1979). However, Coffa's discussion is some-what problematic on the philosophical side, since he erroneously assimi-lates Weyl's epistemology to that of Quine's holism. But that is an issue for another time. Another discussion of the mathematics involved is in T.A. Ryckman, "Weyl, Reichenbach and the Epistemology of Geometry," *Studies in the History and Philosophy of Science* 25 (6) (1994): 831–870.

47   Weyl, *Raum-Zeit-Materie*, 1st ed. (Berlin: Springer Verlag, 1918), 82.

48   Weyl, *Space-Time-Matter*, 102.

49   Weyl, *Space-Time-Matter*, 102.

50   Weyl once held that "Everything real (*Wirkliche*) that transpires in the world is a manifestation of the world metric. Physical concepts are none other than those of geometry." "*Reine Infinitesimal geometrie*," *Mathematische Zeitschrift* 2 (1918): 385.

51   Weyl, *Space-Time-Matter*, 147. (Browse translation modified.)

52   Weyl, *Space-Time-Matter*, 147.

53   Van Dalen, "Four Letters from Edmund Husserl to Hermann Weyl," 11.

54   See Weyl's discussion of the infinitesimal space about a point from the perspective of group theory. *Space-Time-Matter*, 138–148.

55   Weyl, *Erkenntnis und Bessinung*, in his *Gesammelte Abhandlungen*, 4 vols. (Berlin, Heidelberg, New York: Springer Verlag, 1968), 3: 632. This view illustrates Weyl's debt to Hilbert's *Foundations of Geometry*. In *Erkenntnis und Bessinung*, Weyl states that Hilbert's text shook his Kantianism to the ground and opened him to a new way of understanding geometry. Hilbert, in the opening pages, insists that constructing axioms and exploring their interrelations (for Euclidean geometry) is "equivalent to the logical analy-sis of our perception of space." *Foundations of Geometry*, trans. L. Unger from the 10th German edition, rev. and enlarged P. Bernays (Illinois Open Court, 1987), 2. This perception is essentially that of 'local space.'

56   See Ryckman's paper for more on this.

57   Van Dalen, "Four Letters from Edmund Husserl to Hermann Weyl," 11.

58   Weyl, *Space-Time-Matter*, 148.

59   This is not surprising since Weyl once claimed to be raised a "strict Cantorian dogmatist." "Draft for a lecture at the Bicentennial Conference," taken from Skuli Sigurdsson, *Hermann Weyl, Mathematics and Physics, 1900–1927* (PhD diss., Harvard University, 1992), 39. Weyl's early (1910) ontological views are in his "*Uber die Definitionen der mathematischen Grundbegriffe*," in his *Gessamelte Abhandlungen*, 2: 299–304. Here Weyl commits himself to the (ideal) existence of lines, planes and points and holds that we can directly intuit *individual* numbers.

60   In his final paper, Weyl even jettisons the independent existence of the

natural numbers. They are simply "the mind's own free creations." "Axiomatic Versus Constructive Procedures in Mathematics," *The Mathematics Intelligencer* 7 (4) (1985): 17. This article was written shortly before his death in 1953.

61   Stephen Toulmin, "Review of *Philosophy of Mathematics and Natural Science*," *Philosophical Review* 59 (1952): 386.

62   Weyl, "*Erkenntnis und Besinnung*," 155.

63   Weyl, "*Erkenntnis und Besinnung*," 160.

64   Toulmin, "Review of *Philosophy of Mathematics and Natural Science*," 385.

CHAPTER EIGHT

John L. Bell

HERMANN WEYL'S LATER PHILOSOPHICAL VIEWS: HIS DIVERGENCE FROM HUSSERL

———

In what seems to have been his last paper, *Insight and Reflection* (1954), Hermann Weyl provides an illuminating sketch of his intellectual development, and describes the principal influences – scientific and philosophical – exerted on him in the course of his career as a mathematician. Of the latter the most important in the earlier stages was Husserl's phenomenology. In Weyl's work of 1918–22 we find much evidence of the great influence Husserl's ideas had on Weyl's philosophical outlook – one need merely glance through the pages of *Space-Time-Matter* or *The Continuum* to see it. Witness, for example, the following passages from the former:

> Expressed as a general principle, this means that the real world, and every one of its constituents, are, and can only be, given as intentional objects of acts of consciousness. The immediate data which I receive are the experiences of consciousness in just the form in which I receive them ... we may say that in a sensation an object, for example, is actually physically present for me – to whom that sensation relates – in a manner known to everyone, yet, since it is characteristic, it cannot be described more fully.[1]

and:

———

the datum of consciousness is the starting point at which we must place ourselves if we are to understand the absolute meaning of, as well as the right to, the supposition of reality ... 'Pure consciousness' is the seat of what is philosophically *a priori*.[2]

But a reading of *Insight and Reflection* shows Weyl to have moved away from the phenomenology which, as he remarks, "led me out of positivism once more to a freer outlook on the world." This divergence can in fact already be detected in Weyl's *The Open World* of 1932, in which, while granting that

The beginning of all philosophical thought is the realization that the perceptual world is but an image, a vision, a phenomenon of our consciousness; our consciousness does not directly grasp a transcendental real world which is as it appears...The postulation of the real ego, of the thou and of the world, is a metaphysical matter, not judgment, but an act of acknowledgment and belief.

He continues:

It was an error of idealism to assume that the phenomena of consciousness guarantee the reality of the ego in an essentially different and somehow more certain manner than the reality of the external world; in the transition from consciousness to reality the ego, the thou and the world rise into existence indissolubly connected and, as it were, at one stroke.[3]

I submit that Weyl's use of the term "idealism" here is intended to include Husserl's phenomenology, since in *Insight and Reflection* Weyl remarks, in connection with Fichte's philosophy, that "Metaphysical idealism, toward which Husserl's phenomenology was then shyly groping, here received its most candid and strongest expression."[4]

In *Insight and Reflection* Weyl describes Husserl as "an adversary of the psychologism which prevailed at the turn of the century," who went on to develop:

the method of phenomenology, whose goal it was to capture the phenomena in their essential being – purely as they yield themselves apart from all genetical and other theories in their encounter with our consciousness. This quintessential examination unfolded to him a far broader field of evidently *a priori* insights than the twelve principles which Kant had posited as the constituting foundation of the world of experience.[5]

Weyl quotes a number of passages from Husserl's *Ideas*, which he calls his "great work of 1922." But some of Weyl's comments on these passages have a somewhat critical tenor. For example, Weyl says:

To point up the antithesis between an accidental, factual law of nature and a necessary law of being, Husserl cites the following two statements: 'All bodies are heavy' and 'All bodies have spatial extent.' Perhaps he is right, but one senses even in this first example how uncertain generally stated epistemological distinctions become as soon as one descends from generality to specific concrete applications.[6]

He presents his own position on this issue by quoting *Space-Time-Matter*:

The investigations about space that have been conducted [here] seem to me a good example of the kind of the kind of analysis of the modes of existence which is the object of Husserl's phenomenological philosophy, an example that is typical of cases in which we are concerned with non-immanent modes. The historical development of the problem of space teaches how difficult it is for us human beings entangled in external reality to reach a definite conclusion. ... Certainly, once the true point of view has been adopted reason becomes flooded with light, and it recognizes and appreciates what is of itself intelligible to it. Nevertheless, although reason was, so to speak, always conscious of this point of view in the whole development of the problem, it had not the power to penetrate it in one flash. This reproach must be directed at the impatience of those philosophers who believe that, on the basis of a single act of exemplary concentration, they are able to give an adequate description of being. In principle they are right, yet from the point of view of human nature, how utterly they are wrong! The example of space is at the same time most instructive with regard to the particular question of phenomenology that appears to me the decisive one: To what extent does the limitation of those aspects of being which are finally revealed to consciousness express an innate structure of what is given, and to what extent is this a mere matter of convention?[7]

He goes on to say:

Einstein's development of the general theory of relativity, and of the law of gravity which holds in the theory's framework, is a most striking confirmation of this method which combines experience based on experiments, philosophical analysis of existence, and mathematical construction. Reflection on the meaning of the concept of motion was important for Einstein, but only in such a combination did it prove fruitful.[8]

From this passage I think it may be inferred that Weyl had come to hold the view that the ultimate secrets of Being cannot be arrived at by philosophical reflection alone.

Weyl next turns to what he identifies as the central theme in Husserl's work, namely:

the relationship between the immanent consciousness, the pure ego from which all actions emanate, and the real psychophysical world, upon whose objects these acts are intentionally directed.[9]

Weyl characterizes Husserl's view of space as an object for the ego as follows:

Concerning space as an object, Husserl says that, with all its transcendence, it is something that is perceived and given in material irrefutability to our senses. Sensory data, 'shaded off' in various ways within the concrete unity of this perception and enlivened by comprehension, fulfil in this manner their representative 'function;' in other words, they constitute in unison with this quickened comprehension what we recognize as 'appearances of' color, form, etc.[10]

However, Weyl quickly questions this account of the matter:

I do not find it easy to agree with this. At any rate, one cannot disavow that the particular manner in which, through this function of inspiration, an identifiable object is placed before me, is guided by a great number of earlier experiences ... the theoretical-symbolic construction, through which physics attempts to comprehend the transcendental content behind the observations, is far from inclined to stop with this corporeally manifested identity. I should, therefore, say that Husserl describes but one of the levels which has to be passed in the endeavour through which the external world is constituted.[11]

Later Weyl appears to be somewhat uncomfortable with Husserl's epistemological idealism:

Concerning the antithesis of experience and object, Husserl claims no more than merely phenomenal existence for the transcendental as it is given in its various shadings, in opposition to the absolute existence of the immanent; i.e., the certitude of the immanent in contrast to the uncertainty of the transcendental perception. The thesis of the world in its accidental arbitrariness thus stands face to face with the thesis of the pure I and the I–life which is indispensable and, for better or worse, unquestionable. 'Between awareness and reality there yawns a veritable chasm of meaning,' he says. 'Immanent existence has the meaning of absolute being which *'nulla re indiget ad existendum'*; on the other hand the world of the transcendental *'res'* is completely dependent on awareness – dependent, moreover, not just on being logically thinkable but on actual awareness.[12]

This brings Weyl to the enigma of personal identity, a problem to which he ascribes paramount importance:

> Here finally arises in its full seriousness the metaphysical question concerning the relation between the one pure I of immanent consciousness and the particular lost human being which I find myself to be in a world full of people like me (for example, during the afternoon rush hour on Fifth Avenue in New York). Husserl does not say much more about it than that "only through experience of the relationship to the body does awareness take on psychological reality in man or animal."[13]

In this connection it is worth quoting what Weyl had to say concerning this issue in his *Address on the Unity of Knowledge*, delivered not long before.

> It is time now to point out the limits of science. The riddle posed by the double nature of the ego certainly lies beyond those limits. On the one hand, I am a real individual man ... carrying out real and physical and psychical acts, one among many. On the other hand, I am 'vision' open to reason, a self-penetrating light, immanent sense-giving consciousness, or however you may call it, and as such unique. Therefore I can say to myself both: 'I think, I am real and conditioned' as well as 'I think, and in my thinking I am free.' More clearly than in the acts of volition the decisive point in the problem of freedom comes out, as Descartes remarked, in the theoretical acts. Take for instance the statement $2 + 2 = 4$; not by blind natural causality, but because I see that $2 + 2 = 4$ does this judgment as a real psychic act form itself in me, and do my lips form these words; two and two make four. Reality or the realm of Being is not closed, but open toward Meaning in the ego, where Meaning and Being are merged in indissoluble union – though science will never tell us how. We do not see through the real origin of freedom.
>   And yet, nothing is more familiar and disclosed to me than this mysterious 'marriage of life and darkness,' of self-transparent consciousness and real being that I am myself. The access is my knowledge of myself from within, by which I am aware of my own acts of

perception, thought, volition, feeling and doing, in a manner entirely different from the theoretical knowledge that represents the 'parallel' cerebral process in symbols. The inner awareness of myself is the basis for the more or less intimate understanding of my fellow-men, whom I acknowledge as beings of my own kind. Granted that I do not know of their consciousness in the same manner as my own, nevertheless my 'interpretative' understanding of it is apprehension of indisputable accuracy. As hermeneutic interpretation it is as characteristic for the historical, as symbolic construction is for the natural, sciences. Its illumining light not only falls on my fellow-men; it also reaches, though with ever-increasing dimness and incertitude, deep into the animal kingdom. Kant's narrow opinion that we can feel compassion, but cannot share joy, with other living creatures, is justly ridiculed by Albert Schweitzer who asks: 'Did he ever see a thirsty ox coming home from the fields drink?' It is idle to disparage this hold on nature 'from within: as anthropomorphic and elevate the objectivity of theoretical construction, though one must admit that understanding, for the very reason that it is concrete and full, lacks the freedom of the "hollow symbol."' Both roads run, as it were, in opposite directions: what is darkest for theory, man, is the most luminous for the understanding from within; and to the elementary inorganic processes, that are most easily approachable by theory, interpretation finds no access whatsoever.[14]

Returning to *Insight and Reflection*, Weyl goes on to compare Husserl's position with that of Fichte, a philosopher whose views Weyl says also had a pronounced influence on him. Although Weyl claims to find "preposterous" the actual details of what he calls Fichte's "constructivism," according to which the world is a necessary construction of the ego, nevertheless we find him asserting that: in the antithesis of constructivism and phenomenology, my sympathies lie entirely on [Fichte's] side.[15] But he quickly adds:

yet how a constructive procedure which finally leads to the symbolic representation of the world, not *a priori*, but rather with continual reference to experience, can really be carried out, is best shown by physics – above all in its two most advanced stages: the theory of relativity and quantum mechanics.[16]

Soon afterwards Weyl introduces a geometric analogy that, he believes,

will be helpful in clarifying the problem with which Fichte and Husserl are struggling, namely, to bridge the gap between immanent consciousness which, according to Heidegger's terminology, is ever-mine, and the concrete man that I am, who was born of a mother and who will die.[17]

In this analogy objects, subjects, and the appearance of an object to a subject are correlated respectively with *points on a plane, (barycentric), coordinate systems in the plane,* and *coordinates of a point with respect to a such a coordinate system.*

In Weyl's analogy, a coordinate system S consists of the vertices of a fixed nondegenerate triangle T; each point p in the plane determined by T is assigned a triple of numbers summing to 1 – its *barycentric coordinates* relative to S – representing the magnitudes of masses of total weight 1 which, placed at the vertices of T, have centre of gravity at p. Thus objects, i.e., points, and subjects i.e., coordinate systems or triples of points belong to the same 'sphere of reality.' On the other hand, the *appearances* of an object to a subject, i.e., triples of numbers, lie, Weyl asserts, in a different sphere, that of *numbers.* These *number–appearances,* as Weyl calls them, correspond to the experiences of a subject, or of pure consciousness.

From the standpoint of naïve realism the points (objects) simply exist as such, but Weyl indicates the possibility of constructing geometry (which under the analogy corresponds to external reality) solely in terms of number–appearances, so representing the world in terms of the experiences of pure consciousness, that is, from the standpoint of idealism. Thus suppose that we are given a coordinate system S. Regarded as a subject or 'consciousness,' from its perspective a point or object now corresponds to what was originally an appearance of an object, that is, a triple of numbers summing to 1; and, analogously, any coordinate system SN (that is, another subject or 'consciousness') corresponds to three such triples determined by the vertices of a nondegenerate triangle. Each point or object p may now be *identified* with its coordinates relative to S. The coordinates of p relative to any other coordinate system SN can be determined by a straightforward algebraic transformation: these coordinates represent the *appearance* of the object corresponding to p to the subject represented by SN. Now these coordinates will, in general, *differ* from those assigned to p by our given coordinate system S, and will in fact coincide for all p if and only if SN is what is termed by Weyl the *absolute* coordinate system consisting of the three triples (1,0,0), (0,1,0), (0,0,1), that is, the coordinate system which corresponds to S itself. Thus, for this coordinate system, 'object' and 'appearance' coincide, which leads Weyl to term it the *Absolute I*.[18]

Weyl points out that this argument takes place entirely within the realm of numbers, that is, for the purposes of the analogy, the *immanent consciousness*. In order to do justice to the claim of objectivity that all 'I's are equivalent, he suggests that only such numerical relations are to be declared of interest as remain unchanged under passage from an 'absolute' to an arbitrary coordinate system, that is, those which are invariant under arbitrary linear coordinate transformations. According to Weyl,

> this analogy makes it understandable why the unique sense-giving I, when viewed objectively, i.e., from the standpoint of invariance, can appear as just one subject among many of its kind.[19]

Then Weyl adds an intriguing parenthetical observation:

> Incidentally, a number of Husserl's theses become demonstrably false when translated into the context of the analogy – something which, it appears to me, gives serious cause for suspecting them.[20]

Unfortunately, we are not told precisely which of Husserl's theses are the "suspect" ones.

Weyl goes on to emphasize:

> Beyond this, it is expected of me that I recognize the other I – the you – not only by observing in my thought the abstract norm of invariance or objectivity, but absolutely: you are for you, once again, what I am for myself: not just an existing but a conscious carrier of the world of appearances.[21]

This recognition of the *Thou*, according to Weyl, can be presented within his geometric analogy only if it is furnished with a purely *axiomatic* formulation. In taking this step Weyl sees a third viewpoint emerging in addition to that of realism and idealism, namely, a *transcendentalism* which "postulates a transcendental reality but is satisfied with modelling it in symbols." [22]

But Weyl, ever-sensitive to the claims of subjectivity, hastens to point out that this scheme by no means resolves the enigma of selfhood. In this connection he refers to Leibniz's attempt to resolve the conflict between human freedom and divine predestination by having God select for existence, on the grounds of sufficient reason, certain beings, such as Judas and St. Peter, whose nature thereafter determines their entire history. Concerning this solution Weyl remarks:

> [it] may be objectively adequate, but it is shattered by the desperate cry of Judas: Why did I have to be Judas! The impossibility of an objective formulation to this question strikes home, and no answer in the form of an objective insight can be given. Knowledge cannot bring the light that is I into coincidence with the murky, erring human being that is cast out into an individual fate.[23]

Weyl's divergence from pure phenomenology is made evident by the passage immediately following, which shows him to have come to embrace a kind of theological existentialism:

At this point, perhaps, it becomes plain that the entire problem has been formulated up to now, and especially by Husserl, in a theoretically too one-sided fashion. In order to discover itself as intelligence, the I must pass, according to Descartes, through radical doubt, and, according to Kierkegaard, through radical despair in order to discover itself as existence. Passing through doubt, we push through to knowledge about the real world, transcendentally given to immanent consciousness. In the opposite direction, however – not in that of the created works but rather in that of the origin – lies the transcendence of God, flowing from whence the light of consciousness – its very origin a mystery to itself – comprehends itself in self-illumination, split and spanned between subject and object, between meaning and being.[24]

Weyl says that from the late works of Fichte he moved to the teachings of Meister Eckhart, whom he calls the "deepest of the Occidental mystics," the originality of whose basic religious experience cannot be doubted:

It is the inflow of divinity into the roots of the soul which he describes with the image of the birth of the 'Son' or of the 'Word' through God the Father. In turning its back on the manifold of existence, the soul must not only find its way back to this arch-image, but must break through it to the godhead that lives in impenetrable silence.[25]

It was through the reading of Eckhart that Weyl "finally found for himself the entrance to the religious world."[26] But he admits that his metaphysical-religious speculations never achieved full clarity, adding that "this may perhaps also be due to the nature of the matter."[27]

In his later years, Weyl says:

I did not remain unaffected either by the great revolution which quantum physics brought about in natural sciences, or by existentialist philosophy, which grew up in the horrible disintegration of our era. The first of these casts a new light on the relation of the perceiving subject to the object; at the center of the latter, we find neither a pure I nor God, but man in his historical existence, committing himself in terms of his existence.[28]

During his long philosophical voyage Weyl stopped at several ports of call: in his youth, Kantianism and positivism; then phenomenological idealism; and finally a kind of theological existentialism. But apart from his brief flirtation with positivism, itself, as he says, the result of a disenchantment with Kant's "bondage to Euclidean geometry," Weyl's philosophical orientation was in its essence idealistic: he cleaved always to the primacy of intuition that he had first learned from Kant, and to the centrality of the individual consciousness that he first absorbed from Fichte and Husserl. But while he continued to admire Husserl's philosophy, I infer from his remarks in *Insight and Reflection* that he came to regard it as lacking in two essential respects: first, it failed to give due recognition to the transcendental external world, with which Weyl, in his capacity as a natural scientist, was concerned; and secondly, and perhaps in Weyl's view even more seriously, that it failed to deal adequately with the enigma of selfhood: the fact that I am the person I am. Grappling with the first problem led Weyl to emphasize the essential importance of symbolic construction in grasping transcendental external reality, a position which brought him close to Cassirer in certain respects; while the second seems to have led him to existentialism and even to religious mysticism.

Notes

1. Hermann Weyl, *Space-Time-Matter*, originally published as *Raum-Zeit-Materie* (Berlin: Springer Verlag, 1918), tr. Henry L. Brose (New York: Dover, 1950), 4.
2. Weyl, *Space-Time-Matter*, 5.
3. Weyl, *The Open World: Three Lectures on the Metaphysical Implications of Science* (Yale University Press, 1932), 26–7.

4   Weyl, "Insight and Reflection," lecture delivered at the University of Lausanne, Switzerland, May 1954. Translated from German original in *Studia Philosophica* 15 (1955), in *The Spirit and Uses of the Mathematical Sciences,* ed. T.L. Saaty and F.J. Weyl, 281–301 (New York: McGraw Hill, 1969), 285.

5   Weyl, "Insight and Reflection," 285.

6   Weyl, "Insight and Reflection," 288.

7   Weyl, "Insight and Reflection," 290.

8   Weyl, "Insight and Reflection," 292.

9   Weyl, "Insight and Reflection," 293.

10 Weyl, "Insight and Reflection," 293.

11 Weyl, "Insight and Reflection," 294.

12 Weyl, "Insight and Reflection," 296.

13 Weyl, "Insight and Reflection," 298.

14 Weyl, *Address on the Unity of Knowledge,* Columbia University Bicentennial Celebration, 1954. Reprinted in *Gesammelte Abhandlungen Hermann Weyl,* ed. K. Chandrasekharan, 4 vols. (Berlin: Springer Verlag, 1968), 4: 626.

15 Weyl, "Insight and Reflection," 299.

16 Weyl, "Insight and Reflection," 299.

17 Weyl, "Insight and Reflection," 299.

18 This phrase Weyl derives from Fichte, whom he quotes as follows: "The I demands that it comprise all reality and fill up infinity. This demand is based, as a matter of necessity, on the idea of the infinite I; this is the absolute I (which is not the I given in real awareness)." Weyl, "Insight and Reflection," 297.

19 Weyl, "Insight and Reflection," 299.

20 Weyl, "Insight and Reflection," 299.

21 Weyl, "Insight and Reflection," 299.

22 Weyl, "Insight and Reflection," 299.

23 Weyl, "Insight and Reflection," 300.

24 Weyl, "Insight and Reflection," 300.

25 Weyl, "Insight and Reflection," 300.

26 Weyl, "Insight and Reflection," 299.

27 Weyl, "Insight and Reflection," 299.

28 Weyl, "Insight and Reflection," 301.

PART III

PHENOMENOLOGY, THE SCIENCES, AND COMMUNITY

CHAPTER NINE

Pierre Kerszberg

FROM THE LIFEWORLD TO THE EXACT SCIENCES AND BACK

1. PHILOSOPHY SUMMONS SCIENCE

Ever since the rise of the modern exact sciences, it has become more and more clear that the scientific mind is not bound by an exhaustive understanding of its own doings. The fact is that, while discovering the inner structure or the nexus of relations pertaining to an object, science ignores the paths that led to this structure or these relations; but these paths lie precisely at the basis of the ontological ground of the object. Husserl reflected on this legacy of the scientific revolution when he argued that "it is not always natural science that speaks when natural scientists are speaking."[1] The realm of intelligibility that exceeds science's capacity for understanding its own doings is occupied by epistemology. Or rather: it *ought* to be so, because for the most part contemporary epistemology has contented itself with the task of completing science's realm of intelligibility, without acknowledging that another realm of sense is thereby opened up. What exactly is this new realm of sense, if it is issued from science and yet irreducible to the operations of science? The question forms the background of what Husserl referred to in terms of the crisis of the European sciences. The modern sciences have no need for a ground, whether metaphysical or epistemological; they function by themselves in perfect independence from ideal crutches. Yet the supreme philosophical task arises from the realization that, by relinquishing the ground, the sciences leave wide open the reflective field concerned with their meaning. The task is to provide them with a ground, even though they themselves – as far as they can give an account of themselves – have nothing to do with it and will continue to be deaf to it in all foreseeable futures. In sum, the crisis that appeared in the wake of modern science is ultimately a philosophical, not a scientific crisis. How much longer can science continue along its path without being damaged by this paradox of a crisis of meaning which seemingly does not affect its progress?

The conceptual upheavals brought about by the modern sciences of nature could well be gauged against a series of radical uncertainties: namely, at the basis of all scientific concepts and theories we now find undecidedness. Thus, whether space is absolute (Newton) or relational (Leibniz), mechanics goes its own way and remains Newtonian throughout; the same holds for the concept of time. Over and above this pair 'absolute/ relational,' other examples can be found, and it could even be argued that the exact sciences have progressed by virtue of increasing basic irresolutions, such as the continuous versus the discontinuous, the finite versus the infinite, the local versus the global, etc. These indecisions have the peculiarity of being harmless with respect to the content and the development of the scientific theories themselves; they pertain to what is referred to as their 'foundations.' This is perhaps one of the most intriguing paradoxes of modern science: the very progress of science leads gradually to a complete separation of its foundation from its content, as if science had to settle accounts only with itself. Indifferent to its origin, science is essentially preoccupied by its aim, which is the establishment of a unity among the laws of nature. In this way, science finds itself in the uncomfortable position of having to topple one of two radically divergent attitudes: either pure and simple paralysis or endless (and hence blind) renewal of meaning. How are we to orient ourselves in a realm of sense where undecidedness is the only rule? We do not know yet how and where science meets up with philosophical reflection over its sense; yet the philosophical project inherited from Husserlian phenomenology argues that somehow science can be called on to do so. Our task is double: i) identify the point of encounter; ii) understand its nature and its implications.

2. WIDENING SCIENCE

The object proper to science could still be taken as an *indication* of how the reflection over sense should proceed. Historically, this kind of reflection is transcendental reflection on science; it begins with Kant. Ignoring the threat of paralysis, Kant opts for revisiting the renewal of meaning – as exemplified in the scientific revolution of the seventeenth century – by twisting it in a well – defined direction. Kant believed that such a swelling of sense was tantamount to its widening. In *Critique of Pure Reason*, it is argued that every act of constituting objective experience is grounded by the pure *a priori* forms of sensibility, namely, space and time. That is why *Critique* is not, properly speaking, a theory of scientific knowledge (in the sense of attempting to justify its claims transcendentally); rather, it uses well-established scientific propositions as a springboard to assert claims of general scope regarding the possibility of every possible cognition. This becomes quite evident in the principle of pure understanding which deals with quality (the "Anticipations of Perceptions"), where Kant explains what he means by a transcendental proof. Such a proof, Kant argues, reaches the level at which nothing is to be explained, which is why it can be understood in terms of the phenomenological intelligibility of the scientific object.[2] In the case of the principle under consideration, the alleged explanation is related to how space is filled with matter. To be sure, a transcendental proof does not explain anything, but it has 'the merit at least of freeing the understanding,' a merit which in this case applies to how we must think about the phenomenon of the filling of spaces in a non-dogmatic manner. The physicist explains the difference in the quantity of various kinds of matter in bodies that have the same volume by means of the supposition that the real in space is everywhere homogeneous and its volume is empty in varying degrees. Even though this supposition is not verifiable empirically, and is therefore to be counted as a dogmatic position, the physicist, in suggesting such an explanation, claims to be faithful to his own notion of the freedom of mind, since his motivation is that patient and arduous research will eventually lead to the desired confirmation in experience. But by virtue of his own ideal of freedom, the transcendental philosopher urges another supposition against the physicist's: namely, that matter fills space continuously by degrees so that there is no point of space where matter is not present, if only in the smallest possible degree. Kant is well aware that the latter supposition is not meant to assert that this is what actually occurs in real bodies. The point of setting the transcendental critique against the dogmatic explanation of the physicist is to provide a *widening* of the domain of natural science beyond that which science can or must think by its own means.

Since it is usually argued that post-Kantian science has incorporated premises of philosophical character into its own propositions, the question arises whether science has in fact proceeded in the broadening direction advocated by transcendental reflection. The example of the special theory of relativity, which was first developed by Einstein in 1905, suffices for our purposes. In this theory, it first appears that the laws connecting observers in relative motion express a reflective return to the very conditions of the possibility of physical theory. Indeed, the invariance of the laws with respect to the Lorentz group of transformations derives from the employment of two primitive terms, space and time, which are expressed in terms of the measurable behaviour of clocks and rigid bodies; in pre-relativistic physics, absolute space and absolute time were defined abstractly as metaphysical entities unaffected by the practical limitations inherent in any operation of measurement. Is this return to the concrete limitations of possible experience truly 'reflective' in Kant's sense of the term? The answer must definitely be negative, since special relativity forgoes explaining physically the length contraction and the time dilation that result from the measurement of length and time intervals separating inertial observers; both phenomena (contraction and dilation) depend solely on the choice of inertial frame of reference, and they can be interpreted as mere 'appearances' resulting from this choice. In other words, relativistic physics does not comply with the sacrosanct equation that lies at the basis of Kant's principles of pure understanding: namely, that the conditions of possibility of experience be identical with the conditions of possibility of the *objects* of experience. The breakdown of this equation leads to something new.

3. SCIENCE SEPARATES FROM THE LIFEWORLD

In his reflections on the nature of the physical world as described by contemporary physics, Eddington argued that the most significant implication of the scientific revolution that occurred early in the twentieth century is that, from now on, science deals with a world of *shadows*.[3] These shadows are the mathematical symbols. To be sure, there is a real that is correlated to these symbols, but how this correlation actually works is left in the dark. The real no longer exists independently of the observer, in the realm of the in-itself, like a permanent and immutable substance. Kant had already overcome the old metaphysical concept of substance by converting it into a methodological principle for the intelligibility of empirical reality. A theory like quantum mechanics goes perhaps a step further: the real exists only inasmuch as, and as long as, a measurement indicates it. The operation of measurement leads to a phenomenal manifestation and occurs within an indivisible ensemble: the physical system and the experimental device; however, the registered results lend themselves to statistical regularities only. The fuzziness of the phenomenon is the price to pay for including the actions of the observer in the actions of nature. That is why the world has now split in two. First, there is the world of the always present, unquestioned real; second, there is the world of symbols, in which the abstract part of the real has become independent from its counterpart in actually experienced reality. The symbolic world is enigmatically connected to the world from which we come by the operation of measurement. For the mediation of measurement to become intelligible as a faithful mouthpiece of the one, single being of nature, bridges connecting the two worlds should be found. That is, one should imagine one world out of which the other two could be derived. According to Eddington, these bridges still existed in classical physics, but the representation of a unique world is no longer allowed. As he puts it:

> It is true that the whole scientific inquiry starts from the familiar world and in the end it must return to the familiar world; but the part of the journey over which the physicist has charge is in foreign territory.[4]

The mind of the physicist is like an alchemist's mind that transmutes symbols: by confronting them with experience, he gives them meaning. For example, the agitation of molecules becomes the experience of warmth. On the one hand, the physicist runs the risk of falling back into the illusions of classical science if he leaves the foreign territory in order to seek the substance again. But on the other hand – and this would be the great lesson of recent advances in science – the transmuted world propels us into a realm of new meanings; the latter have their proper horizon and effectiveness. The preoccupation with a return to origins – the familiar world – world has become inessential. But is it even possible?

The widening of science by philosophical critique has thus finally come down to the separation between the ideal world of symbols and the concrete world of possible experience. This is particularly clear in quantum mechanics, where the precedence of the measurement device over the phenomena to be measured is really disturbing. Experimental results are interpreted in accordance with a theoretical structure thanks to which the limits to predictions can be set. But the inclusion of the experimental device in the very series of such results leads to the impossibility of complying with essential criteria related to the existence of phenomena such as their reproducibility, their individuality, or their spatio-temporal identifiability. To be sure, the apparatus is defined by a certain project in accordance with which it performs certain tasks. The project derives from a well-defined horizon of sense, but how could this horizon be maintained in light of such results, indeed how could it be recognized at all in the results? One does not see what kind of schematization could ensure the transition from the project included in the apparatus to the measurements themselves, since the latter seem to violate the limits of the initial project.

As regards contemporary natural science, its progress lies undeniably in its emancipation from the notion of an actually existing observer, that is, an observer equipped with consciousness associated with the full range of sensory powers. This emancipation begins with the special theory of relativity, in which the causal structure of space-time has been derived from the measurements made by an 'ideal' observer, that is, a device reduced to the recording of time-intervals separating the reception of light signals. From a Kantian viewpoint, it could be argued that, because of this reduction of sensibility to its minimal capacities, the ideal observer records the events of the world thanks to a kind of intellectual intuition, whereby a sensible manifold of objects is directly ordered by means of intuition alone, in conformity with the 'being' of these objects. In *Critique* Kant admits this possibility: nature is the sum of the objects given to the human senses or to some other kind of intuition.[5] Thus, physical science founded upon intellectual intuition is perfectly conceivable. But in that case, human reason would no longer have to distinguish itself from a superior understanding; what Kant takes to be the necessarily unsatisfied requirement of all life, i.e. the drive by inward needs towards the ideal world of metaphysics, would then be satisfied by natural science alone. As we know all too well, the current discourse of natural science is almost indistinguishable from what used to be referred to as metaphysics. Suffice it to think about the cosmological speculations on the earliest phases of the universe; in these speculations, relativity and quantum mechanics co-operate so that phenomena and events that are below the theoretical threshold of observability could be made intelligible.

What distinguishes the Husserlian from the Kantian critique is that, according to Husserl, the sense organs of the living subjects could be quite different from what we find them to be, yet the essential condition to reach the intelligibility of the natural world is that the senses should, at the very least, allow for a common understanding (*gemeinsame Verständigung*) between subjects, without which it would be impossible to constitute a common nature as it appears in some way to the senses (*gemeinsame Natur als erscheinende*).[6] This is Husserl's project: enlarge the realm of the sensible *a priori* (as understood by Kant) to the point that it has enough resources to support the weight of the ever growing domain of empirical experience as revealed by the natural sciences.[7] But how is this possible if the empirical domain were to grow so far that it falls into the non-recoverable world of symbols? In the wake of contemporary natural science, the task of transcendental philosophy of nature thus distinguishes itself inasmuch as it has to *lower* the ideal components of science to the level of the lifeworld and its own *a priori* structures. In Eddington's account, the journey back to the familiar world appeared inessential, or even impossible, but this familiar world was not questioned for its own sake. By contrast, all theories asserting the autonomy of the "true" physical thing, Husserl writes:

> are possible only as long as one avoids seriously fixing one's eyes on, and scientifically exploring, the sense of what is given in the thing (*Dinggegebenem*) and, therefore, of 'any thing whatever,' a sense implicit in experience's own *essence*.[8]

To be sure, one might be willing to accept the view that the thing determined in accordance with physics is not foreign to what appears sensuously. But in order to make the journey back to the lifeworld not only possible, but in fact necessary, Husserl adopts the more radical stance according to which the physical thing "makes itself known originally in it and, more particularly, *a priori* (for indefeasible eidetic reasons) *only* in it."[9]

Husserl's strategy has thus been clarified. Let us recall that Husserl's aim is to provide a ground for the sciences, where in fact none is required or even desired. At first it would seem that, in the absence of reliable support from the object as it appears to the senses, all that philosophy can do is to start from the *idea* of ground, as if it could be posited by analogy with an ideal limit; it would be up to the object to conform to it. But then the sensuously-apprehended object is treated as nothing more than a misleading sensory impression, or as the image or sign that stands for the true determinations of the object. There must be a way in which the object – as it is apprehended sensuously – continues to present itself as an obstacle on the royal road toward symbolic science. Here the object is no longer indication, as it was for Kant, but rather *resistance*. As such, if the idea of ground is not to exhaust itself in a vain battle against its volatile enemy, it is something to be circumvented in such a way that the resistance exerted by the object impresses the idea itself. The resistance of the object comes essentially from the appearance inasmuch as there is something irreducible to it: its ability of being given as 'itself in person,' i.e. its *Leibhaftigkeit* or living corporeality.

4. LIFEWORLD SEPARATES FROM SCIENCE

Starting from sensible appearance, pure consciousness has to wrench itself from natural consciousness. Indeed, only such a consciousness, as it withdraws from the stream of natural experiences, can outline the sense of the transcendent object in relation to the knowing subject. But this pure consciousness, which posits the thing as itself in person, immediately collides with the finite limits of this appearance.[10] Indeed the appearance offers itself through a series of adumbrations, so that, from the standpoint of our closed consciousness, and within the limits of finite appearances given through partial adumbrations, no object of nature can be given in an adequate manner. Correlatively, the universal (or eidetic structure) that provides the sense of the judgments of experience is also affected by the limitations of adumbrations, so that it could not be said to be unequivocal. Therefore, if there is a sense in which a necessarily harmonious concatenation of appearances is thinkable, i.e. a system in which the ever more exact determination of a natural being would be protected from any possible contradiction, it involves transition to an *infinite* consciousness.[11] Such a consciousness would prescribe ideally the indefinite process of continuous appearances. Thus, as a regulative ideal, only the Kantian Idea of a transcendent object can reach the level of perfectly adequate givenness. But Husserl's interpretation of this Idea goes well beyond the regulative function assigned to it by Kant. Indeed, the Idea does not abrogate absolute contingency, but on the contrary it is meant to maintain it by combining it with absolute necessity.

> This continuum is determined more precisely as infinite on all sides, consisting of appearances in all its phases of the same determinable X so ordered in its concatenations and so determined with respect to the essential contents that any of its *lines* yields, in its continuous sense, a harmonious concatenation (which itself is to be designated as a unity of mobile appearances) in which the X, given always as one and the same, is more precisely and never 'otherwise' continuously-harmoniously determined.[12]

Starting from *any* line of experience that lies open before us, the subjective consciousness of concordant appearances will nevertheless not fail to reach the self-same object, i.e., the object would be the same if the process had started from another line. So, with the Idea, an *a priori* basis or eidetic type is provided, in which are sketched the infinite, always open-ended, possibilities of the future stream of experience.[13] The aim of science is to provide an absolute determination of objects or natural processes in accordance with the Idea of such objects or processes. Consequently, to the ordered series of adumbrations provided by the perception of an object, is collated another ordered series of adumbrations of the very idea of this object. Do these two series of adumbrations complement each other harmoniously?

There exists an order of the world that differs from the essential connections apprehended by the pure constitutive consciousness. This is the morphological order, which is the intentional correlate of the factual connections pertaining to lived experiences. Whereas the former lends itself to exact and universal mathematical laws of nature, the latter belongs to the domain of the sciences of classification and description. The morphological order is itself an 'appearance' built upon the essential order of the exact natural laws.[14] Evidently this exactness/universality does not prescribe anything with regard to the existence of facts, such as the 'appearance' of our earth with its creatures at such or such moment in the history of the solar system. Hence, there follows what Husserl calls "a marvellous teleology," which reflects the divergence between essential rationality and rationality as it is realized in fact. This divergence motivates the employment of the method of transcendental reduction, whereby the question is raised "about the ground for the now-emerging factualness of the corresponding constitutive consciousness," that is to say, the constitution corresponding to the teleology revealed by factual existence. As regards pure consciousness, the whole range of possible judgments about this teleology – for example, that the universe evolves in such a way that rational beings emerge at a certain time in order to take control of nature – will be affected by the very same contingency that characterizes the facts that these judgments are supposed to explain. The teleological judgment does not and cannot get rid of the contingency of that upon which the judgment bears.

Not the fact as such, but the fact as source of endlessly increasing value-possibilities and value-actualities forces the question into one about the 'ground' – which naturally does not have the sense of a physical-causal reason.[15]

So far we have traced the relations of fore-understanding that lead from the merely perceived object to the idea of the object: perception and ideation by adumbrations. At the very moment when the fact is finally to present itself as ground, it is no longer perceivable as sheer fact. Rather, it is perceived as endowed with value; it carries an order that escapes the grip of pure consciousness available originally. Even though it is still immersed in the world, pure consciousness might well be compelled to pass its constitutive power on to another consciousness, for which the 'marvellous teleology' will present itself as a series of adumbrations with its own order, not as an absolutely contingent series. This other constitutive power could well be that of a 'divine' consciousness. Now, as the scientific revolution of the seventeenth century has taught us, such a divine consciousness, if it can be conceptualized at all, happens to be no different than mathematical consciousness, which certainly brings it closer to human capacities. That is why, as Husserl argues, God is a transcendency to be bracketed out, much like all natural transcendencies.

If phenomenology is to align itself with the critical function of transcendental reflection, then it must be viewed as the investigation of those acts of consciousness leading to the fatal moment when the power of pure consciousness gives out, when it leaves room for another absolute power. Besides the constitutive acts, what are the acts of consciousness leading up to this fatality? On the one hand, since the time of Galileo, physical reality is defined independently of supposedly absolute acts of consciousness; that is, reality is dealt with as if it could be regarded as a world valid in itself (*an-sich-Sein*). On the other hand, the phenomenological constitution of worldly objects leads to a view of physical reality which treats it as directed toward pure consciousness, i.e. as a *Sein-für* in relation to such a consciousness. The task is to examine how the world in itself, whatever it may be (Husserl marks it out as an "empty X"), can still be made intelligible from the standpoint of pure consciousness, assuming however that the lesson taught by the modern exact sciences is irrevocable, that is, an unbridgeable gap will always continue to separate the mode of being of each of these two worlds.

In the spatial perception of the *individual thing*, what I aim at exceeds what the thing invites me to see. A sphere of absence or invisibility surrounds the thing and constitutes it. Now, the mechanism of this perception reverses our apprehension of the *world*. Indeed, if the world is to be regarded as a kind of being, this is in a very special sense, namely, "with such uniqueness that the plural makes no sense when applied to it."[16] The world is the presupposition of every plurality, but only inasmuch as it moulds the perception of the individual so that the void that exceeds it can be taken in. The world is ground for as long as it does not act as ground, i.e., as long as it withdraws from the things that it conditions. That is why in the natural perception by adumbrations, where eidetic variations follow their course within the limits set by the thing itself, the series of adumbrations ends up by being unable to satisfy the requirement of immediate apprehension of the thing. In the final analysis, the thing itself has been self-destroyed. Now, in section 47 of *Ideen* I, Husserl asks himself if the thing, as viewed from the standpoint of physical science, is not in fact superior to the thing as intuited in perception, so that the latter is subordinated to the former. As a matter of fact, the order of human experience necessarily leads to theoretical interest – science – in which one conjectures the existence of the "truth of physics"[17] underneath the data of intuition. Does this mean that, when necessity constrains reason in this way, the experience of the world has reached the point that intuition has no part in it? In order to see whether this is the case, the necessity in question must be subjected to an eidetic variation of a radical kind: this course of experience, Husserl points out: "might be different." In this case, the correlate of empirical consciousness would be destroyed; in fact, the objectivity of things would be unable to fix itself in the things themselves. Obviously, this unique eidetic variation is nothing other than the *epoché*. In this new attitude, consciousness comes face to face with its own responsibilities with regard to the world, since all belief in the existence of things is now put out of play. What is it that consciousness discovers? The destruction of *things* gives way to a new range of eidetic forms, i.e., nothing less than a plurality of *worlds*. Husserl writes:

> the correlate of our factual experience, called 'the actual world', is one special case among a multitude of possible worlds and non-worlds which, for their part, are nothing else but the correlates of essentially possible variants of the idea, 'an experiencing consciousness', with more or less orderly concatenations of experience.[18]

The multiplicity of worlds, which includes non-worlds (adumbrations in chaos), comes to the rescue of the solitary consciousness, cut off from natural things, incapable of making contact with the one single actual world which it aims at; in going beyond these particular things, it defines "an undetermined but *determinable* horizon," in which all that exists really without being yet actually experienced "can become given."[19] Incapable of preserving to the end its own demand for immediate apprehension, the perception by adumbrations falls into the chaos of adumbrations, but in turn, this chaos opens it up to non-worlds. To be sure, the perception of a thing given in intuition might always lead to the realization that, in fact, this thing was not in or of the world in which the perception began; reality as constituted by pure consciousness can never be more than presumed reality. Nevertheless, according to Husserl, it is precisely by virtue of this risk that the consciousness of the genesis of the given (in the perception by adumbrations) is able to withdraw from the single given world-horizon; in so doing, the given itself now originates from the chaos – a multiple horizon – in which it was not yet the 'itself in person' which it is about to become. The multiple horizon provides a deepening of the sense of the given, leading back to its own genesis, when pure consciousness faces the limits of its constituting activity. From now on, when viewed from the multiple horizon, the possibility and necessity of things ensure the potential – if not the actual – fulfilment of intuition; they form a set of prescriptions in which the arbitrariness of the eidetic variations has been eradicated. Put briefly: the multiple horizon has rescued the natural perception from self-destruction.

Now, the separation between familiar experience and the symbolic world of natural science is to be traced back to the foregoing separation between the characters of singular perception and the characters of the multiple horizon. According to Husserl, the starting point of all cognition is not the neutral object, apprehended as such, but the intentional object that results from a certain delimitation of the originally fuzzy intuitive field of phenomena.[20] As it turns out, as Husserl argues in *Ideas II*, the delimitation specific to scientific cognition implies the taking into consideration of the living body. Only in this way can the separation be properly understood.

5. THE RETURN TO/OF THE LIFEWORLD

The problem of the constitution of natural objects, as it arises from the living body, has to face the representation of physical reality provided by the physical sciences. Thus, heat is understood in terms of the agitation of molecules, but the subjective sensation of heat is declared to be confined to the living body; the latter is a subjective impression that is completely separated from the objective phenomenon itself. Obviously, if this complete separation were to be maintained, the process of constitution would never come to completion, but would rather bump into the 'thing itself' as an impassable wall. Yet, according to phenomenology, the thing 'is' only in its possible or actual correlation to a well-determined phase of constitution. One solution runs like this:[21] the constitution of the object depends on the actual orientation (*Einstellung*) of the body in different phases of consciousness; this orientation could be directed to the phenomenon in a variety of ways, including whether the phenomenon is an agitation of molecules, or whether it is a merely subjective impression. There would be two planes of adumbrations, as it were: each such plane would yield the total object as either 'physical' or 'perceptual.' However, consider the perception of a certain colour, which from the standpoint of physics is made intelligible in terms of the emission of a well-defined wavelength of light. From the double standpoint of the thing itself and its perception by a subject, the emission of a certain wavelength of light and the sensation of colour happen *simultaneously*. The thing would be simultaneously seen as coloured and yet be uncoloured in its 'being.' Therefore, the continuity of perceptual adumbrations, whereby the living subject is able to anticipate the normal course of a phenomenon in the flux of consciousness (either by confirmation or by spontaneously corrected invalidation), is broken. Is a mere change of orientation of the body enough to justify this rupture? If, in order to constitute the physical object, the living subject must at some point dispossess itself from its own prerogatives, then, once more, it seems difficult to escape the undesirable conclusion that some mysterious 'natural' causality operates instantaneously in the world, so that the final product of the constitution is an 'illusion' (*Schein*) or a purely subjective image that stands for an undeterminable 'X.'

Physics proceeds by means of 'jumps' from 'lower' to 'higher' levels. Generally speaking, at any time of its development, the reality of a lower level becomes the 'appearance' of a higher level. This is already visible in the constitution of a body. Thus, a solid body, as dealt with in purely geometrical terms, is originally made of the sum of all relative viewpoints that we have of it; each such appearance is a perspective relative to its absolute determination. At the end of the process, there will still be a certain relation between the objective conditions of the thing itself and the most immediate data of consciousness.[22] But how can something that does not appear be taken as the condition of that which appears? Independently of any level reached by physics, nature as it appears *already* includes the non-appearing as one of its conditions: these are the *modes* through which the thing appears. Yet these modalities, even though they do not appear, do not quite escape the relativity of daily experience, since the state of the subject can be either 'normal' or 'pathological' (alterations of the perceptual field, etc). Perhaps the distinction between the 'seeming' and the 'effectively real' is never possible in the final analysis. Nevertheless, the thing of physics can be seen as an abstraction of a higher level only inasmuch as it reaches complete detachment from all relativity. Thus, according to theoretical physics, to see a thing with eyes capable of occupying a range of wavelengths different from habitual perception is not a case of abnormal perception.

In order to reach the level of the fully constituted physical thing, Husserl finds it necessary to distinguish between different planes of adumbrations, not in terms of perceptual versus physical objects, but in terms of the transition for a living subject from solitude to community; the latter embodies the multiple horizons opened up by science and gives it its proper phenomenological foundation.[23] First, consider a solitary subject immersed in a 'solipsistic world,' surrounded by material things only. How does one go from relative experiences (perceptions by potentially self-destroyed adumbrations) to the identical substratum? Two stages must be distinguished. 1) The thing presents itself to me in intuitive experience with all its optimal features, both geometrical and sensible: this is the thing 'itself' or 'in itself.' 2) Nevertheless, the transition to a higher level (the thing as physical thing) is 're-quired' (*erfordert*) because the thing itself is correlated in a certain way, and is therefore relative to, the living body in its *Leiblichkeit*. Here, in the logical and geometrical construction, the sensible phenomenon departs from its qualitative features. What does this relativity mean? How does the geometrical become autonomous as the thing itself is correlated to the living body? The transition would never occur if the living body did not require it. Indeed, the body begins to distinguish between real and unreal phenomena or events in the external world when the normal correlation between its own organs of perception begins to fail; in the event of such a lapse of one organ with respect to the others, the need for retrieving a balance implies that "the distinction between *Schein* and *Wirklichkeit* is always given;"[24] the geometrical properties gain their independence from the sensible qualities as a result of this balance.[24] This is how the living subject of experience is able, by means of its own forces, to 'expel' from within itself the character of the thing as it is in itself; the thing in physical science does not result from the mere elimination of the relative in favour of the absolute (as if the secondary qualities could be simply brushed away so that only the primary qualities remain). At this point, it would thus seem that the truth of physics (the separation between the geometrical and the sensible) succeeds in overcoming the relativity of perceptions by adumbrations (which ultimately, as we have seen, destroy themselves). But now the snag comes from the corporeal subject itself.

The subject, as it is dealt with in the exact sciences, is an ideal subject, that is, ideally unaffected by such accidental changes in its perceptual makeup. Therefore, if perceptions were to be viewed as always 'normal,' which is the tacit assumption in these sciences, then, since the aforementioned distinction would not count, an objective nature could never be arrived at for such a subject. From the standpoint of phenomenology, just as the adequate perception of a spatial thing is impossible, the alleged truth of physics turns out to be a contradiction. Now, the solipsistic subject is not yet a real subject either; in reality it is actually surrounded by other subjects. Let us imagine what happens when suddenly, for the first time, other human beings are present, with whom I can communicate (*mich veständigen*).[25] At first, the result is devastating, since this means again a return to the original relativity of all experience. By virtue of the primary relativity of all experience, most of the propositions concerning things that I have been able to establish in solitude would *not* be confirmed by others. In the first communication between my organ of perception and a foreign organ, it looks as if my own organ has failed. Other people will take my construction of reality as hallucination; I cannot but conclude that my physical world returns to the level of *Schein* and that the attempt to extricate myself from it failed. In the transition from the solipsistic to the intersubjective world, the physical thing that was first constructed is self-destroyed, much like in the perception by adumbrations where the intuitive thing is self-destroyed.

At this point, therefore, the power of the alleged 'physical' truth (the truth that the exact sciences supposedly reach independently of phenomenological constitution) gives out; it is of no assistance, since it would merely transpose the multiplicity of other bodies into the multiplicity of other worlds and non-worlds. Is there still a common horizon in which these worlds and non-worlds would eventually co-operate in some way to ensure the genesis of the 'given' in corporeal self-hood? But in the actual communication between living subjects, the first 'given' that presents itself is nothing other than my own proper body, which is now objectified by virtue of this very communication with other such bodies. The awareness of my own body becoming a fully 'human' body, when set in conjunction to other similar bodies, shows that there was something in the solipsistic experience of things that must have been incomplete. Now we see that the self-destruction of the intuitive thing came too soon, as it were. To be sure, the perception by adumbrations *anticipates* the full validation (*Ausweisung*) of the thing, but it does not *require* it (in fact, Husserl identifies the full validation with a Kantian Idea of Reason). Intersubjective experience, as it discloses the existence of one humanity in which all subjects recognize themselves in their living corporeality, leads to a new kind of interrogation of the world: what is it in the thing that is still hidden (an as yet unfulfilled intention) over and above all possible adumbrations, and requires validation? If this hidden dimension can be discovered and identified, the threat posed to the thing as constituted in the solipsistic experience of the world (namely: that as a sensible and intuitive thing it will vanish as pure *Schein*) will be neutralized: with respect to objective reality (*objektive Wirklichkeit*), the intuitive thing will be an *Erscheinung*, not a *Schein*.[26] The answer, as we shall see now, is that the notion of 'one determined human-ity' does not and cannot overlap with 'one undetermined but determinable horizon.'

Since, in intersubjective awareness, the appearances are now relative to such or such living body, the mode whereby they are given, which was itself not appearing in the solipsistic experi-ence, is now something that one can seek to see. The mode is no longer relative to one determined subject, and therefore there is an experience of it which is not yet ostensive, but, as Husserl puts it, "requires ostensivity." In response to this demand, spatial form must be distinguished in terms of "appearing form" (*erscheinende Gestalt*) and form (or magnitude) itself (*Gestalt selbst*). The latter, too, can be said to 'appear,' but it does so from the standpoint of an intuition that is higher than that of the former, because it emerges as one and the same over and above every change of place and/or modification of empathy.[27] In this way, when space itself, as a mode of experience, becomes something intuitively given 'in person,' it immediately allies itself with the properties of motion of the spatial bodies; motion, in turn, is inconceivable outside the flow of time as one and the same time. Eventually, by means of the common, spatio-temporal deter-mination, the thing can be viewed as "the rule of its possible appearances."[28]

Intersubjective experience reveals that there was more in the solipsistic experience (i.e. my own proper body as something objective) than could be gathered by its own means. The discov-ery of this hidden dimension in my body reminds me that the other realities are variations of the self-same world. We now see that that which goes beyond adumbrations, and is yet still at-tached to the thing, is the fact that all things are located in one and the same space, one and the same time; they are objective forms by virtue of being single forms. This definite sameness in space–time differs from the indeterminate world–horizon in prescientific cognition. Both the world–horizon and the single space–time form are 'one,' but in a different sense. While the former is the undetermined (non-sensible) and yet determinable stock of all potential intuitions, and contains the subjective conditions of genesis of the sensibly given, in the latter, the distinction between sensible and non-sensible is a matter of organization, not genesis, of the sensible.

Thus Husserl has shown that the deepening of the sense of the thing from the standpoint of physics can be accomplished only by deepening the sense of the genesis of the thing in eidetic intuition: this deepening takes us beyond the constituting power of pure consciousness, into the genesis of the given itself; but this genesis, in turn, presents itself to consciousness as the rule whereby all potential and actual givens insert themselves in the continuum of the self-same nature.

Nevertheless, Husserl's final point is that the thing of physics is the same, whether it is constituted solipsistically or intersubjectively. Of course, in the case of intersubjective constitution, objective space and objective time are included in the final product, whereas neither objective space nor objective time are taken as presuppositions in the constitution of the physical thing by a single subject. The double-sidedness of the process reflects the problematical status of the whole of modern physics. Indeed, in the Newtonian synthesis, objective space and objective time vanish as appearances, they become 'absolutes,' once they can be conceived as the asymptotic limits of relative (measurable) space and time. As their appearance is eliminated, they can be taken as the non-appearing conditions of possibility for the appearance of all spatio-temporal determinations. In positing its absolutes, modern physics returns to the solipsistic world in which the ideal and the sensible were still inextricably interwoven. The advent of relativistic space–time, and then quantum mechanics, have not changed this fact: they have contributed to a better awareness of it. Thus, in Special Relativity, the new absolute (the four-dimensional continuum of space–time) coalesces the two separate absolutes of Newtonian mechanics (absolute space and absolute time), so that the line of demarcation between the ideal and the sensible is buried even more deeply in the formalism of the theory. Witness the problematical status of the constancy of the velocity of light in this theory, which can be interpreted either as a fact or as a stipulation. In quantum mechanics, such a fundamental principle as Heisenberg's principle of indeterminacy transposes limits of measurability into intrinsic limits of particle-like or wave-like behaviour of reality, with the consequence that the line of demarcation between the virtual and the actual can no longer be set for once and for all.

6. CONCLUSION: WIDENING THE LIFEWORLD

Husserl has repeatedly emphasized that, over and above the differences between the world–horizon of lived experience and the space–time background of physics, there is yet only one world to which everything belongs: scientific as well as all other cultural and technical activities. The special character of the *epoché* that Husserl has articulated in *Crisis* is to bring the sense of these activities back to a common ground. Arguably, it is not the case that the knowledge claims of science belong to the lifeworld – they are independent of the phenomenological reduction. Rather, Husserl wants to show that the sciences belong to the lifeworld only as a fragment of the cultural forms of humanity. Nevertheless, there remains a sense in which scientific claims themselves remain anchored in the lifeworld: this is related to the evolution of the role of experimentation in natural science. No object of modern natural science can be intelligible outside prior experimental preparation; the purpose of experimentation is to provide a passage from the sensible world to the absolutes (ultimate conditions of possibility) of natural science. The effect of this preparation is to blur the limit at which the object 'in-itself' begins to be intuitively perceptible, or conversely, the limit at which the intuitively perceived object begins to steep in the realm of the in-itself. Every process of preparation is like a reminder of the fact that the specialized objects of physics still belong to the lifeworld – however specialized the latter has itself come to be. One could speak of the preparation–world in science as a segment of the lifeworld; what comes after the preparation propels us into the in-itself, and this latter world is a world that we can 'live' in for as long as it remains meaningfully connected to the original preparation. In quantum mechanics, this connection is conceived in terms of probability, but that does not change anything concerning the fact that the outcome of experiments remains attached to the original preparation – though in this case the attachment is a matter of 'after-thought,' whereas in classical physics it was a matter of 'pre-thinking.'[29] Because of this double but obscure involvement, and even though in quantum mechanics the physicist hits at a more and more narrow segment of the lifeworld, the physical object ultimately cannot make sense independently of the intuitive (whether actual or merely possible) capacities of a living subject; nor can the intuitive world ignore the historical transformation of world objects under the increasing influence of science, so that the *"lebensweltlichen Dingen"* can no longer be seized absolutely for what they are supposed to be. Thus, when Husserl describes the lifeworld in terms of a set of originary evidence to which every constitutive act must ultimately return, he does not ignore the fact that this set is itself defined by purely relative and changing subjective structures; the point is that this relativity must itself be correlated, in substantial respects, to the changing worldviews brought about by science.

Notes

1   Edmund Husserl, *Ideas Pertaining to a Pure Phenomenology and to a Phenomenological Philosophy, First Book*, § 20, tr. F. Kersten (The Hague: Martinus Nijhoff, 1982), 39. See the profound remarks on this subject by E. Levinas "Réflexions sur la technique phénoménologique," in his *En découvrant l'existence avec Husserl et Heidegger* (Paris: J. Vrin, 1974), 116. The starting point of the present paper is found in E. Ströker, "Edmund Husserl's Phenomenology as Foundation of Natural Science," in *The Husserlian Foundations of Science*, ed. L. Hardy, 111–121 (Washington: University Press of America, 1987).

2   Immanuel Kant, *Critique of Pure Reason*, tr. N.K. Smith (London: Macmillan, 1929), A174/B215–6, 206.

3   A.S. Eddington, " Introduction " to *The Nature of the Physical World* (Cambridge: Cambridge University Press, 1928), xi–xix.

4   Eddington, *The Nature of the Physical World*, xv.

5   Kant, *Critique of Pure Reason*, A845/B874, 662.

6   Husserl, *Ideen zu einer Reinen Phänomenologie und Phänomenologischen Philosophie*, M. Biemel ed., Book II (The Hague: Martinus Nijhoff, 1952), 86.

7   Hermann Weyl, *Philosophy of Mathematics and Natural Science* (Princeton: Princeton University Press, 1949), 111.

8   Husserl, *Ideen* I, 118. (Translation modified).

9   Husserl, *Ideen* I, 120.

10 Husserl, *Ideen* I, 331.

11 Husserl, *Ideen* I, 342.

12 Husserl, *Ideen* I, 342.

13 Husserl, *Ideen* I, 343–344.

14 Husserl, *Ideen* I, 133–134.

15 Husserl, *Ideen* I, 342.

16 Husserl, *The Crisis of European Sciences and Transcendental Phenomenology*, tr. David Carr (Evanston IL: Northwestern University Press, 1970), 143.

17 Husserl, *Ideen* I, 105.

18 Husserl, *Ideen* I, 106.

19 Husserl, *Ideen* I, 107.

20 Husserl, *Ideen* I, § 37.

21 See Roman Ingarden, "Husserls Betrachtungen zur Konstitution des physikalischen Dinges," in *La Phénoménologie et les Sciences de la Nature*, *Archives de l'Institut International des Sciences Théoriques* (Brussels: Office International de Librairie, 1965), 13: 36–87.

22 Weyl, *Philosophy of Mathematics and Natural Science*, 113.

23 Husserl's way out of the threat of return to illusion is articulated in the passages of *Ideas* II dealing with the constitution of the physical thing: sections 18d to 18g.

24 Husserl, *Ideen* II, 78.

25 Husserl, *Ideen* II, 79.

26 Husserl, *Ideen* II, 82.

27 Husserl, *Ideen* II, 83.

28 Husserl, *Ideen* II, 86.

29 The emphasis on pre-thinking has been articulated by Heidegger in *What is a Thing?* tr. Barton and Vera Deutsch (Chicago: Regnery, 1969).

CHAPTER TEN

R. Philip Buckley

HUSSERL ON THE COMMUNAL PRAXIS OF SCIENCE

---

1. INTRODUCTION

It is well known that for a long period within the phenomenological tradition itself, there was a tendency to view the *Crisis*-
texts of Husserl's last years as marking a radical shift in his
thought. Major figures such as Gadamer and Merleau-Ponty[1] are
well-known exponents of this view, and even circumspect and
insightful subsequent scholars such as Carr tend to stress the
novelty, for example, of the infusion of history into Husserl's
later philosophy.[2] Some treat this 'novelty' as a reaction to the
historical crisis of the 1930s, and also imply that the proximity
and popularity of Heidegger should not be ignored.[3] To the
contrary, in some of my previous work I have tended to stress the
unity of the Husserlian project from beginning to end, to suggest
that such apparently 'novel' topics as 'history' themselves have a
long genesis in Husserl's work and that the 'idea' of a crisis in
European culture in general is itself best understood by paying
close attention to the manner in which Husserl addresses 'earlier' crises such as foundational questions in mathematics.[4] Of
course, using 'crisis' as a *leitmotiv* for Husserl's thought in its
entirety does not resolve all the tensions in his philosophy. But it
helps us to understand these tensions as belonging to the internal

---

dynamic of his project rather than arising from some unfortunate collision between radically different conceptions of philosophy which occur at incommensurate stages of his life-work.

I will use this unified approach as a hermeneutic key to consider another component of Husserl's thought: his theory of community. As of late, there has been a small but growing interest in Husserl's social philosophy. A cynic might suggest that the reason for this is that Husserl's writings on inter-subjectivity were published in 1973 and that it has taken the normal human being somewhere between 20 and 30 years to digest these three massive volumes of *Husserliana*.[5] A less 'psychologistic' interpretation of this phenomenon is that philosophers have realized that Husserl has an original contribution to make to the debate between radical, liberal individualism, and communitarianism. Whatever the motivation, in works such as Schuhmann's *Husserls Staatsphilosophie*[6] and Hart's *opus magnum* entitled *The Person and the Common Life: Studies in a Husserlian Social Ethics*, there is a movement to take seriously, if not critically, what Husserl has to say about society and the foundation of human community.[7]

This paper aims to supplement this movement by expanding the scope of the focus on Husserl's theory of community. Much of the work on this still under-developed aspect of Husserl's thought centres on the last half of *Husserliana XIV* (e.g. *Gemeingeist I & II*) and *Husserliana XV* and the *Kaizo* articles of 1922–23 found in *Husserliana XXVII*.[8] It is true that Husserl's reflections on community become most explicit in his later works. Nevertheless, a return to certain earlier works and experiences can aid us in gaining insight into these later reflections. In the first part of this presentation, a succinct summary of the main components of Husserl's mature theory of community is given and a major difficulty in this theory is pointed out: namely, a certain tension between individual insight and shared communal insight. In the second part, a suggestion is put forth as to how this tension might be – if not resolved – at least understood by returning to certain aspects of Husserl's earlier work. Finally, the conclusion will stress that even if this tension remains unresolved, Husserl's reflections on community have much to offer which is of contemporary significance.

2. HUSSERL'S UTOPIA

The central feature of Husserl's theory of community is the notion of 'personality of higher-order.' Different sorts of communities ranging from the family to the state are understood as analogous to the individual 'I.' A community is a "multi-headed ... yet connected subjectivity,"[9] and as such a subjectivity it too has a 'personality,' displaying particular tendencies, moods, and traits such as memory – indeed, many of the features usually attributed to individual existence. Nonetheless, the community which possesses a personality *is* of a 'higher-order.' It is different than the individuals who comprise it, and its personal features belong to it 'uniquely.' That is, the personal features of the community are something 'new;' they are not a mere conglomeration of individual features. It is here that a first ambiguity or tension arises in Husserl's theory of community, but I think it is a fruitful tension, or at least, a tension that has to do with Husserl's phenomenological acuity. Husserl is trying to account for the real identity which occurs within various types of communal existence without making community some sort of pre-existent structure which has enveloped individual existence. Conversely, Husserl's notion of 'personality of higher-order' tries to maintain the essential aspects of individual existence, while still showing that something arises out of that individual existence that is truly new and different. Hence, a community is a real entity in its own right, but it cannot exist apart from individuals. Conversely, individuals produce something which at the same time is both completely dependent on those same individuals, but also exceeds them. One is reminded here of the nature of categorial acts, which on the one hand are something truly new, but on the other hand are founded and exist only on the basis of individual acts of perception. It is therefore not surprising that Husserl's treatment of community often mirrors the back and forth hermeneutic that characterizes his description of categorial acts in *Logical Investigations*: a continuous attempt to show how these founded acts are *both* rooted in *and* somehow beyond individual acts of perception.

One positive result of Husserl's constant re-phrasing of the relationship between individual and community is that it yields a description of how an identity is gained *through* a community, without resorting to some metaphysical notion of the whole. It is true, however, that one can uncover within this fine example of phenomenological description a number of prescriptive elements. Though ontologically founded upon the individual, 'communities' *can* dominate the individual and be arranged in a hierarchical and domineering fashion. Husserl describes such an "inauthentic" community as an "imperialist unity of will" wherein individuals are subordinate to, and submit themselves to, a central will.[10] Husserl, who never displayed a particularly subtle view of empirical history, took the medieval Church as the best example of a communal organization which in a certain sense undermines its own foundation: that is, it is a power-organization which negates its origin in the free decision on the part of *individuals* to pursue in common a shared goal.

The 'authentic' community, on the other hand, is one in which each individual, rather than being subordinate to the 'whole,' actually has insight into the whole and how one fits into the whole which the individual has produced. Or put slightly differently, the individual has an insight into the *collective insight* of the community, one knows how one's free activities merge with the free activities of others to produce a collective activity that is something larger than the sum of its parts. But this image of an emergent collective insight based on participation seems to imply at least two problematic presuppositions. First, it requires a high level of individual authenticity, that is, it demands that individuals know what they are doing and why they are doing it – a heavy demand in view of the repetitive and self-forgetful nature of modern technological life. Indeed, since for Husserl (and this is true from his earliest analyses of 'authentic counting' in *Philosophy of Arithmetic* through to his call for recapturing of 'authentic' European culture at the end), authenticity always means producing the originary evidence for what one does and holds, or at the very least, being willing to strive to produce such evidence, to struggle to recollect 'evidence' from the layers of passivity and sedimentation that allow for 'smooth' functioning but also 'thoughtless' functioning – one is left with the impression that the only authentic community is one wherein everyone becomes a 'phenomenologist.' This may well be a cure for a mindlessly functioning community or culture; on the other hand, one might well question both the chances of survival and the appeal of living in a community consisting *only* of phenomenologists. Certainly, the mechanics of how we progress from hard-earned individual insight to individually possessing the communal insight founded upon individual insight remains puzzling.

Husserl seems to be aware of many of these difficulties, and hence provides a concrete example: the community of mathematicians is used to explain this strange feat of simultaneously pursuing an individual activity and realizing that this individual activity is actually a 'part' of a larger activity. This analogy is by no means foreign to Husserl's thought: in *Kaizo* II he makes it clear that philosophy ought to achieve for science in general what mathematics achieves for the natural sciences. So, too, the community of mathematicians is seen as an example of authentic community, one which the community of philosophers and community at large ought to emulate. The authentic community is:

> similar to the way the collective of mathematicians today forms a community of will, insofar as each individual work concerns a science which is a common good, and hence is intended for every other mathematician. In this community of will, the work of each mathematician profits from the work of every other mathematician, and present in everyone is the consciousness of the totality, of a common goal and of the work which ought to be mutually determining and determined. There is a universal bond of wills present

which establishes the unity of will. This occurs without an imperialist organization of will, without a central will in which all single wills are centered and to which all single wills subordinate themselves readily and as whose functionaries all individuals understand themselves. [At this point, Husserl adds the remarkable footnote: "Here we could also speak of a communist unity of wills in contrast to an imperialist unity."] Here, there is consciousness of the communal goal, of the common good to be pursued, of an encompassing will of which all know themselves to be functionaries, but as free (a freedom which does not need to be renounced) and not subordinated functionaries. (It is otherwise in specific organizations of will such as Academies, etc.)[11]

What can we make of this example? Given that the notion of authenticity itself first appears within Husserl's early reflections on number and his distinction between authentic counting and calculation, and that these early reflections occurred within a framework which linked Husserl to the 'community' of mathematicians, it is not implausible to seek further clarification of this example, and thereby Husserl's ideal of authentic community, by 'questioning-back' into the origin of Husserl's philosophy itself.

3. HUSSERL IN GÖTTINGEN

In some sketchy notes from one of Husserl's earliest lectures in 1887, we find a most powerful statement condemning overspecialization in science and the type of inauthentic thinking which such specialization engenders. Husserl states:

the complete researcher who strives to be a complete human being as well should never lose sight of the relation of his or her science to the more general and higher epistemic goals of humanity. Professional restriction to a single field is necessary; but it is reproachable to become fully absorbed in such a field. And the researcher must appear even more reproachable, who is indifferent even to the more general questions which concern the foundation of his or her science, as well as its value and place in the realm of human knowledge in general.[12]

Here we have a most noble statement about science as a collective activity, one wherein scientists must grasp the meaning of their individual research within the framework of science in general and locate their singular efforts within the striving of all humankind for knowledge and meaning. One might think that a direct line can be drawn between such an ideal and the notion of authentic community from the 1920s which I have just outlined.

There is, however, a further aspect of Husserl's early thought which in some ways works against this noble ideal. In the sentence immediately preceding the previous citation, Husserl admits that specialization – though an 'evil' – may in fact be necessary for progress in science. Husserl's earliest encounter with scientific work already reveals this understanding of 'necessary specialization.' It is well-known that Husserl's early thought was deeply influenced by the effort of Weierstrass to provide a solid foundation for arithmetic by means of a rigorous development of the real number system. Though we know that this early experience can be seen as foreshadowing Husserl's lifelong concern with 'foundation–work,'[13] we also know that Husserl parted company with his mentor on 'who' should accomplish this foundational enterprise. This sort of ground work was, according to Husserl, properly a *philosophical task*.[14] But here we already have, in a nascent stage, the tension which is supposed to be overcome in authentic community as exemplified in none other than the community of mathematicians. That is, we see a group of 'specialists' at work which is not in a position to understand and ground the meaning of its own collective activity. This must somehow be done by a group of other 'specialists or 'authorities,' in this case, philosophers, who are interested not merely in certain singular achievements of mathematics nor simply in its proper functioning, but in its meaning and in its foundation.

There is an ambiguity in Husserl's approach to scientific activity that I think he never fully overcame. On the one hand he accepted the advantages of specialization, and at times almost implies that for science to make progress it must be focused, and hence forgetful of broader concerns and deeper meanings. On the other hand, such forgetfulness of meaning is 'reproachable.' With regard to community, this ambiguity can perhaps be re-phrased in terms of an atomistic, functionalistic vision of society where at best a very few understand how all the parts fit to-gether, versus a wholistic, organic, participatory model where members fit together and form a goal-oriented unit wherein they comprehend how their individual efforts contribute to their com-mon goal. It is clearly this latter notion which is at work in Husserl's vision of authentic community in the 1920s – exempli-fied by none other than the community of mathematicians, who, at this nascent stage, seem to represent specialization and its forgetfulness. What might have caused such a shift?

There are many plausible answers to this question, but I suggest that Husserl's experience at Göttingen to a great extent helped him to re-focus his original ideal, and come to see math-ematicians as perhaps the best example of individuals who grasp the meaning of their own work and its relation to the collective activity of science. By Husserl's 'experience,' I mean more than the internal development of his own thought in this period; I refer to his participation not only in the formal debates within mathematics at Göttingen, but even the institutional structure of the university at that time.

What did Husserl encounter in Göttingen? To begin with, we know that Husserl interacted with a group of mathematicians who were completely devoted to foundational problems. Though this did not preclude the possibility of a *Grundlagenstreit*, the fact remains that these thinkers were absolutely committed to the establishment of a solid foundation for mathematics, and thereby laying a solid foundation for all of physical science. Concurrent with this vision of foundation work is also the fact that Hilbert was a strong opponent of narrow-minded specialization. Overcoming narrow specialization did not entail, however, that one had to be a specialist in many fields! Rather, it meant for Hilbert that one must oppose the constantly recurring idealized picture of the isolated scholar and the value of his or her quiet, reclusive scholarship. To the contrary, the researcher must take an active interest in the effect one's work has on others. In Hilbert's case, this led to an almost prophetic activism, and he felt obliged to promote vigorously what he called the "mission of mathematics."[15]

To have an effect on others requires that one must be in contact with them. Here, one of the more remarkable features of Göttingen was its interdisciplinary nature: philosophy, logic, mathematics, and physics were institutionally linked and considered as one faculty. This institutional fact undoubtedly contributed to the breadth of vision in people such as Hilbert, and perhaps could be said to have legitimized their 'philosophical' aspirations. Indeed, thinkers in Hilbert's mode saw themselves as 'philosophers.' This is clearly stated in his goal to make Göttingen, as he says in a letter to Becker, the "central location for systematic philosophy," and as formulated elsewhere, to build it into "the primary centre for philosophy in Germany."[16]

For Hilbert, this mission consisted concretely in giving to pure theory the leadership role in science. Göttingen was, as Heelan has suggested, the "theoretical cockpit which waged war against the cockpit of pure experiment."[17] The aim of science became the construction of an axiomatic deductive model wherein the chief characteristics were to be completeness, independence, and consistency. That is, the model must be based on independent axioms that do not produce contradictions and be complete in that the truth or falsity of any statement formulated in the system can be determined by calculation. Husserl shared with Hilbert an interest in axiomatic model theory and understood well the importance of such forms for natural science. Indeed, Husserl showed himself somewhat of a prophet regarding the impact which Hilbert and the circle of mathematicians around him were going to have on physics in the early part of this century when he stated that "just as the old quantitative mathematics was the big instrument of research in the natural sciences ... so the new formal mathematics will accomplish much more ... ."[18]

It would be false, of course, to suggest that Husserl was in complete agreement with his Göttingen colleagues. Indeed, many of the essential features of the type of 'Galilean' science critiqued in Crisis are present in Hilbert's ideas. Husserl, while granting that theory was a tremendous deductive aid and lent a precision, a power to computation that signaled a real advance for science, refused to equate the quest for theory with the quest for truth.[19] Nevertheless, Heelan correctly points out that our tendency to focus on Husserl's critique of Galilean science can blind us to the fact that Husserl did not seek to abandon it, but to add to it what was necessary to make it true, complete science – the search for truth.[20] Keeping this in mind, it can be asserted that while Husserl rejects certain claims about the cognitive status of 'theory,' he is able to accept the general spirit of Hilbert's programme. No wonder then, that he would later write so excitedly to Hermann Weyl – that "Hilbert is developing a new foundation for mathematics – wholly in the phenomenological spirit."[21] Husserl would view such a development as paralleling his own efforts: preserving the scientific ethos so beautifully enunciated by Hilbert, and channeling it onto the correct path to truth.

In his famous lecture from 1900 before the Second International Congress of Mathematicians in Paris, Hilbert summarized his hope for mathematical science as a collective activity, and in doing so remarkably prefigures Husserl's statement from Kaizo. Hilbert concluded:

> the question is urged upon us today whether mathematics is doomed to the fate of the other sciences that have split up into separate branches, whose representatives hardly understand one another and whose connection becomes ever more loose. I do not believe this, nor wish it. Mathematical science is in my view an indivisible whole, an organism whose vitality is conditioned upon the connection of its parts. For with all the variety of mathematical knowledge, we are still conscious of the similarity of logical devices, the

relationship of the *ideas* in mathematics as a whole and the numerous analogies in different departments.[22]

What becomes clear here is that to understand the 'whole' as a totality of group activity does not necessarily mean that one has to understand the *content* of the whole. This suggestion, that although there exists a variety of knowledge, there is also a formal similarity between these varieties, helps us to understand what Husserl might be expecting from a community at large that is meant to mirror the community of mathematicians. It is not as if everybody must accomplish the same task or fully grasp the meaning and significance of someone else's task, but everyone must complete his or her own task with a sense of its relation to the whole and with the 'form' of phenomenology: having insight into what one is doing, taking responsibility for it, and justifying all of one's position-takings on the basis of evidence. While not every person must conduct intricate analyses of transcendental consciousness, every individual must 'do *as* phenomenologists do': work towards the goal of a rational existence at the individual level and through this, at the communal level. Such working in the 'same way,' though in different areas, can only enhance a collective movement to rational existence. As Hilbert says a little later on in his Paris lecture, "the sharper the tools for individual research, the better the researcher understands the whole."[23] For Husserl, the more successful the individual is at exercising the form of phenomenological life, the more insight the individual will have into the collective *telos* of humanity to be rational.

4. CONCLUSION

Husserl's ideal of authentic communal life which mirrors Hilbert's idealized vision of the activity of mathematicians offers a number of positive aspects worth mentioning by means of conclusion. Not only is Husserl's ideal community extremely anti-authoritarian and strongly opposed to mindless functioning (even if functionalism should be the most efficient way for a society to work), but its formal quality means that it is extremely egalitarian and participatory. Everybody can participate through individual actions in a form of collective rational existence, everybody can act in the *style* of the phenomenologist, and because of this everybody can gain some insight into how their individual activity contributes to the collective activity.

This openness does come with demands, some of which may be too heavy for most to meet. Clearly, many are content with a life of 'activity in passivity.' Moreover, underlying all of what has been said is the fact that it requires a great deal of *work*. Hilbert's community of mathematicians is one engaged in a common work, and Husserl's authentic community calls for individuals to work hard cultivating their own 'field of values.' Husserl's 'work ethic' can sometimes be a little over-bearing. After all, he was occasionally critical of that other 'Göttingen Circle' around him, those bright young students, including Edith Stein, who would frequent coffee-houses, carry on free-flowing discussions, and listen to guest-speakers like Max Scheler: this was all somewhat superficial to Husserl and no substitute for the labour of phenomenological investigations. Authenticity takes work, and authentic community arises only out of individuals working in similar fashion in the various fields of human endeavour. This is reflective of Husserl's extreme voluntarism, and when it comes to the differing forms of human culture, art for example, it may be unacceptable. Nonetheless, the notion that authentic community can only arise out of hard work highlights Husserl's belief that a community which rests on unreflected traditions, pure functionality, or at worst, merely racial determinations, can never be truly authentic.

It is the *activity* of individuals which founds the collective activity, not some sort of passively received inheritance. Still, though it is authentic individual activity which lies at the base of community, this activity can never be one of mere self-interest. Individuals must, as Husserl says, see the fruits of their activity as a *common good*. To go back to our example of the community of mathematicians – nobody *owns* the Pythagorean theorem. Work within an authentic community is public, not reclusive, and if owned by anybody, it is owned by the public. Husserl's work ethic can be taken in many ways as antithetical to the work ethic of capitalism. Though he stresses individual autonomy, Husserl moves away from an ethics of self-interest. The individual must always be willing and able to justify his or her work, but that justification is 'owed' to the public. A crucial part of that justification and responsibility for one's work is to highlight how the work contributes to the common good. It is thus somewhat surprising that we find in Husserl, who is so often characterized as 'Cartesian' and somewhat distant from everyday life, the basis of a political philosophy which calls incessantly for a move away from self-interest to a higher collective interest.

Notes

1   Maurice Merleau-Ponty, *Phénoménologie de la perception* (Paris: Gallimard, 1945), 61.
2   David Carr, "Husserl's Crisis and the Problem of History," in Carr, *Interpreting Husserl: Critical and Comparative Studies* (Dordrecht/ Boston/ Lancaster: Martinus Nijhoff, 1987), 71–73. Elsewhere, Carr does emphasize the continuity of Husserl's thought *vis-à-vis* history; see his *Phenomenology and the Problem of History* (Evanston, IL: Northwestern University Press, 1974), 66–67.
3   See Paul Ricoeur, "Husserl and the Sense of History," in P. Ricoeur, *Husserl: An Analysis of His Phenomenology* (Evanston IL: Northwestern University Press, 1967), 144.
4   See R. Philip Buckley, *Husserl, Heidegger and the Crisis of Philosophical Responsibility* (Dordrecht: Kluwer Academic, 1992); parts of the present article have appeared as "Husserl's Göttingen Years and the Genesis of the Idea of Community," in *Reinterpreting the Political: Continental Philosophy and Political Theory*, ed. L. Langsdorf and S. Watson, 39–49 (Albany: SUNY Press, 1998).
5   Edmund Husserl, *Zur Phänomenologie der Intersubjektivität. Texte aus dem Nachlass. Erster Teil (1905–1920), Zweiter Teil (1921–1928), Dritter Teil (1929–*

1935), hrsg. von Iso Kern, *Husserliana* XIII–XV (den Haag: Martinus Nijhoff, 1973).

6 Karl Schuhmann, *Husserls Staatsphilosophie* (Freiburg: Karl Alber Verlag, 1988).

7 James Hart, *The Person and the Common Life: Studies in a Husserlian Social Ethics* (Dordrecht: Kluwer, 1992).

8 Edmund Husserl, *Aufsätze und Vorträge 1922–1937*, hrsg. von T. Nenon und H.R. Sepp, *Husserliana* XXVII (Dordrecht: Kluwer, 1989), 3–122.

9 Husserl, *Aufsätze und Vorträge*, 22.

10 Husserl, *Aufsätze und Vorträge*, 53.

11 Husserl, *Aufsätze und Vorträge*, 53.

12 Husserl-Archive manuscript *K I 28/25a*. Sections of this manuscript are published in Edmund Husserl, *Studien zur Arithmetik und Geometrie*, hrsg. von I. Strohmeyer, *Husserliana*, vol. 21 (den Haag: Martinus Nijhoff, 1983). See 231 for the above citation, which is probably from Husserl's 1887–1888 lectures entitled, *"Einleitung in die Erkenntnistheorie und Metaphysik."*

13 Husserl-Archive manuscript *B II 23/8a*.

14 J. Philip Miller, *Numbers in Presence and Absence: A Study of Husserl's Philosophy of Mathematics* (The Hague: Martinus Nijhoff, 1982), 6.

15 David Rowe, "Klein, Hilbert, and the Göttingen Mathematical Tradition," *Osiris* 5 (1987): 187.

16 Volker Peckhaus, *Hilbertprogramm und Kritische Philosophie: Das Göttinger Modell interdisziplinärer Zusammenarbeit zwischen Mathematik und Philosophie* (Göttingen: Vandenhoeck and Ruprecht, 1990), 223.

17 Patrick Heelan, "Husserl, Hilbert, and the Critique of Galilean Science," in *Edmund Husserl and the Phenomenological Tradition*, ed. R. Sokolowski (Washington DC: CUA Press, 1988), 160.

18 Husserl, *Philosophie der Arithmetik*, hrsg. von L. Eley, *Husserliana* XII (den Haag: Martinus Nijhoff, 1970), 432.

19 Husserl, *Philosophie der Arithmetik*, 343.

20 See Heelan, "Husserl, Hilbert, and the Critique of Galilean Science."

21 D. van Dalen, "Four Letters from Edmund Husserl to Hermann Weyl," *Husserl Studies* 1 (1984): 7.

22 Constance Reid, *Hilbert* (New York and Heidelberg: Springer Verlag, 1970), 83.

23 Reid, *Hilbert*, 84.

INDEX